300

Current Topics in Microbiology and Immunology

E. J. H. J. Wiertz and M. Kikkert (Eds.)

Dislocation and Degradation of Proteins from the Endoplasmic Reticulum

With 19 Figures and 3 Tables

Springer

Emmanuel Wiertz, Ph. D.
Marjolein Kikkert, Ph. D.

Leiden University Medical Center
Department of Medical Microbiology
Albinusdreef 2
2333 ZA Leiden
The Netherlands

e-mail: e.j.h.j.wiertz@lumc.nl
m.kikkert@lumc.nl

Cover illustration by Peter M. Deak, Zlatka Kostova, Antje Schäfer and Wolfgang Hilt (chapter 3, this volume)

ISSN 0070-217X
ISBN-10 3-642-42177-6 Springer Berlin Heidelberg New York
ISBN-13 978-3-642-42177-8 Springer Berlin Heidelberg New York

Springer is a part of Springer Science+Business Media
springeronline.com

Softcover re-print of the Hardcover 1st edition 2005

Editor: Simon Rallison, Heidelberg
Desk editor: Anne Clauss, Heidelberg
Production editor: Nadja Kroke, Leipzig
Cover design: design & production GmbH, Heidelberg
Typesetting: LE-TEX Jelonek, Schmidt & Vöckler GbR, Leipzig
Printed on acid-free paper SPIN 11314189 27/3150/YL – 5 4 3 2 1 0

Preface

The present volume of *Current Topics in Microbiology and Immunology* contains seven chapters that illuminate various aspects of a protein's genesis and terminal fate in the endoplasmic reticulum (ER). This area is of immediate medical relevance and has blossomed, to no small extent, because of the study of molecules central to the function of the immune system [immunoglobulins, T cell receptors, major histocompatibility complex (MHC)-encoded products]. Similarly, the clever strategies used by bacteria or viruses to gain a foothold in the host and ensure their continued survival have uncovered altogether new cell biological principles. It is therefore fitting that a special volume be devoted to the interplay between pathways of protein degradation in the ER and a wide variety of pathogens. The concept of quality control emerged with the appreciation that, in the case of multimeric glycoproteins, any unpaired glycoprotein subunit had great difficulties leaving its site of synthesis—the ER—and was destroyed instead. Free immunoglobulin heavy chains were probably the earliest documented example of this kind, and were long known to cause pathology when their accumulation went unchecked.

Increased knowledge of the biosynthetic pathways of glycoproteins allowed the identification of the ER as an important site where such quality control decisions were made. The T cell receptor for antigen, long considered the paradigm of this mode of degradation, led the way in these early explorations. A major puzzle remained: if the ER is indeed the compartment where nascent chains make their first appearance, begin to fold, and assemble with partner subunits, how then is the distinction made between terminally misfolded proteins and those on the way to a functional end-product? Part of the solution, surely, was the appreciation that the synthetic and degradative functions of the ER might be topologically distinct: by exercising quality control in the lumen of the ER, but relegating proteolysis to the cytosol, this paradox might be solved. There is now ample evidence that such separation indeed occurs, although it is unlikely to be the only solution.

It is the study of viruses and bacteria that has helped identify some of these escape hatches from the ER. Intoxication with cholera toxin or the dislocation reaction used by herpes viruses to dispose of class I MHC products are but two examples, dealt with in this volume (chapters by Lord et al. and Kikkert et al.). Especially the latter studies would have been impossible, were it not for the detailed insights into the biosynthesis of glycoproteins in general, and of MHC products in particular. Two chapters in this volume, by Molinari and Sitia and Groothuis and Neefjes, deal with the general and MHC-specific aspects of glycoprotein synthesis, assembly, and intracellular transport. Yeast geneticists have pursued a parallel route: the wonderful toolbox of Sec mutants has been the key ingredient to a dissection of the eukaryotic secretory pathway. Surely the rules that operate in the ER to ensure delivery of the proper products apply across the eukaryotic kingdom. Using genetics, a host of genes were identified in yeast that could be linked to the disposal of misfolded proteins. It is gratifying to see how the studies initiated in yeast converge with the more biochemically inspired experimental systems in higher eukaryotes, again exemplified by three chapters in the present volume, those by McCracken and Brodsky, Wolf and Schäfer, and Bar-Nun. Given the involvement of the cytoplasmic compartment in disposal of unwanted ER proteins, it is perhaps not surprising to see the ubiquitin-proteasome system make its obligatory appearance: the multitude of Ub ligases required for recognition of an extremely heterogeneous set of substrates is only now beginning to be unraveled. A helicopter view of these pathways is provided in the chapter by Kikkert et al. It would be misleading to suggest that there is now uniformity for pathways of quality control in the ER, and a consensus has yet to emerge. What about the unfolding requirements, so as to be able to deal with partially folded proteins? What about the identity of the channels via which proteins escape from a membrane-delimited compartment? What about the routes used by soluble versus membrane proteins? Polytopic membrane proteins? How many different pathways for such escape exist? The continued examination of viruses, bacteria, and their products is required to answer these questions, as pathogens rely on the exploitation of these cellular pathways for their survival. The chapter by Lord et al. provides an overview of our understanding of the routes traveled by cholera toxin. While certain commonalities with ER degradative pathways are apparent, there are also important differences. It stands to reason that the final picture will be even more complex.

It is perhaps equally important to point out which areas have not been covered in this volume, for lack of reliable information. The most successful approaches in the field of protein quality control have relied on the tried and trusted methods of genetics and biochemistry. Many aspects of cellular physiology are under the control of processes that are not template-encoded, and cannot be manipulated simply by introduction of a mutant cDNA or a siRNA vector. Glycosylation, lipid modifications, and other post-translational modifications must surely be superimposed on the basic toolbox of the host and its pathogens. Continued improvements in analytical methodology coupled with the development of new chemistry-based tools to interfere in these processes should be a high priority to illuminate the many different functions of the ER.

Boston, MA, USA, October 2005 *Hidde L. Ploegh*

List of Contents

List of Contributors

(Addresses stated at the beginning of respective chapters)

CTMI (2006) 300:1–15

The Secretory Capacity of a Cell Depends on the Efficiency of Endoplasmic Reticulum-Associated Degradation

M. Molinari[1] (✉) · R. Sitia[2]

[1]Institute for Research in Biomedicine, Via V. Vela 6, 6500 Bellinzona, Switzerland
maurizio.molinari@irb.unisi.ch

[2]Università Vita-Salute SanRaffaele, DiBiT, Via Olgettina 58, 20132 Milan, Italy

Abstract Plasma cells, like other "professional" secretory cells, are capable of secreting thousands of proteins per second. To accomplish this impressive task, they contain a highly developed endoplasmic reticulum (ER), where newly synthesized proteins must fold and assemble to native structures before secretion. Protein biogenesis in the ER is coupled to a tight quality control schedule: aberrant molecules produced upon failure of the folding/oligomerization processes are retained in the ER, and eventually degraded by ER-associated degradation (ERAD) pathways. The activity of the ERAD machinery therefore needs to be adapted to variations in the load of the ER with cargo proteins. If ERAD is insufficient, misfolded proteins accumulate causing ER stress, apoptosis, and ER storage diseases. The capacity of ERAD also critically determines the efficiency of protein secretion. Here we summarize recent findings highlighting the role of ERAD in disease and development, particularly in professional secretory cells.

1 Protein Synthesis and Folding in the Endoplasmic Reticulum

In mammalian cells, proteins that either reside in intracellular compartments such as the ER itself, the Golgi, and endo-lysosomes, or will be secreted (hormones, antibodies, pancreatic enzymes, etc.), or displayed at the plasma membrane (receptors, channels, etc.), are synthesized by ribosomes attached at the cytosolic face of the ER. Nascent chains are translocated at an average rate of three to five amino acids per second into the ER lumen. Here, they undergo folding, assembly, and other posttranslational modifications, with the assistance of a network of resident chaperones and enzymes.

During or immediately after translocation, most secretory proteins are modified by addition of branched oligosaccharides (N-glycans) or formation of disulfide bonds. These covalent modifications increase the general stability of the final protein products: interestingly, they also facilitate their folding and quality control by mediating and timing interactions with ER folding assistants. N-glycans mediate association with calnexin (Cnx) and calreticulin (Crt), two lectin chaperones.

For certain proteins, the formation of non-native disulfides and their subsequent isomerization are essential for proper folding (Jansens et al. 2002). These reactions are catalyzed by oxidoreductases of the protein disulfide isomerase (PDI) superfamily (Ferrari and Soling 1999; Freedman et al. 2002). The oxidoreductase ERp57 forms functional complexes with Cnx and Crt, thereby coupling glycan modifications and disulfide bond modifications for glycoproteins entering the Cnx/Crt cycle (Helenius and Aebi 2004; Trombetta and Parodi 2003). Other chaperone complexes operate in the ER and may intervene sequentially in folding assistance (Molinari and Helenius 2000) or provide alternative pathways for the myriad different polypeptides to be manufactured in this organelle (Ellgaard and Helenius 2003; Sitia and Braakman 2003).

1.1 Folding Is Often Inefficient

Folding efficiency may drop substantially and can be the cause of impaired cell, organ, and/or organism viability when genetic mutations occur, upon unbalanced synthesis of oligomeric complexes subunits, absence of essential cofactors or prosthetic groups, defective protein glycosylation, or impaired control of redox homeostasis (see Table 1). However, even under normal conditions, protein folding may fail to some extent, resulting in the generation of by-products that need to be destroyed to avoid ER constipation. Surprisingly, the folding of heterologous proteins may be faster and more efficient compared to the maturation of certain cell self-proteins. It may take only 10 min for a viral glycoprotein (e.g., influenza virus hemagglutinin) to complete the folding and oligomerization processes and to leave the ER with an efficiency approaching 100%, but several hours before only a fraction of the cystic fibrosis channel (CFTR) becomes native and leaves the ER. About 80% of the wild type CFTR is degraded (Kopito 1999), a fate shared with other cellular proteins and likely reflecting inherent difficulties in the folding processes.

A certain degree of inefficiency in protein folding may actually be essential for our life, providing peptides to be presented in the context of class I molecules for inspection by immune cells. Whatever its physiological meaning, an inherently inefficient folding implies that the higher the production of the protein factory, the higher the generation of waste that needs to be cleared from the folding compartments. The equilibrium needs to be readjusted and

Table 1 Potential causes of ER stress (and related pathologies)

Mutations that retard or preclude folding [cystic fibrosis and many others (Aridor and Balch 1999)]
Defective synthesis of a subunit precluding assembly [heavy chain disease (Witzig and Wahner-Roedler 2002)]
Reduced level of chaperones or folding assistants [e.g., calnexin in combined lipase deficiency, CLD (Briquet-Laugier et al. 1999), in particular tissue-specific ones such as Hsp47 for collagen (Matsuoka et al. 2004)]
Reduced levels of UPR-signaling molecules [Wolkott Rallison syndrome (Harding and Ron 2002)]
Proteasomal insufficiency [polyglutamine diseases (Bossy-Wetzel et al. 2004)]
Absence of essential cofactors [e.g., vitamin C for collagen, scurvy]
Exuberant synthesis of a normal protein [professional secretors, normal and malignant plasma cells, B cell differentiation, myeloma (Ma and Hendershot 2003)]
Imbalanced redox conditions (hypoxia-reperfusion)
Defective glycosylation [congenital disorders of glycosylation CDG (Shang et al. 2002)]

maintained when a resting cell is woken up to become a factory producing high amounts of secretory proteins, for example during transformation of a B lymphocyte into an antibody-secreting cell. The increased production of secretory proteins activates ER resident stress sensors, ultimately strengthening the folding and degradation machineries of the ER.

1.2 Endoplasmic Reticulum Folding and Disease

During development, there are many conditions that weaken the folding efficiency and/or preclude the assembly of a protein. For instance, the early phases of B lymphocyte development are characterized by the synthesis of Ig-H chains in the absence of L chains: a surrogate light chain is produced by B cell progenitors so as to test the capability of newly generated H chains to pair up later with L chains and make a functional antibody. These complex events are intimately connected with signaling pathways controlling allelic exclusion and Ig-gene rearrangements (Melchers et al. 2000; Hendershot and Sitia 2004). On the other hand, professional secretory cells face the problem of producing proteins in massive quantities. How do these cells build and manage such efficient protein factories? This question is particularly relevant in view of the fact that many debilitating human diseases are caused by defective ER protein folding and quality control (Aridor and Balch 1999). Many of these "conformational diseases" represent loss of function conditions, in which a membrane or secreted protein is retained and degraded from the ER. Examples of this type are cystic fibrosis, familial hypercholesterolemia, diabetes mellitus, osteogenesis imperfecta, and retinitis pigmentosa. The common underlying cause is that mutated proteins do not acquire their native structures and are diverted to ERAD. However, there are also many examples of gain-of-function diseases. If disposal is not efficient, aberrant proteins accumulate in or outside cells, triggering severe damage to cells and tissues. An interesting case is the hereditary lung emphysema caused by α1-antitrypsin deficiency (α1-antitrypsin is the principal blood-borne inhibitor of the destructive neutrophil elastase in the lungs). Mutated α1-antitrypsin is not secreted from liver cells and actually accumulates forming intracellular deposits. The loss of function phenotype observed at the level of patient's lungs is therefore accompanied by a gain of toxic function phenotype at the level of the liver (Perlmutter 1996). It is not clear why only a fraction of the patients with lung problems undergo hepatopathy. This observation implies variability in the individual's capability of coping with protein accumulation in ER quality control machineries.

2
Molecular Mechanisms Mediating Endoplasmic Reticulum-Associated Degradation Selectivity and Timing

Figure 1 summarizes the crucial steps underlying ERAD, each of which can become the target of pharmacologic or genetic manipulation.

A terminally misfolded glycoprotein must be:

1. Recognized and extracted from futile folding cycles
2. Disassembled and at least partially unfolded
3. Dislocated across the ER membrane
4. Deglycosylated and ubiquitinated by cytosolic enzymes
5. Degraded by proteasomes or alternative pathways

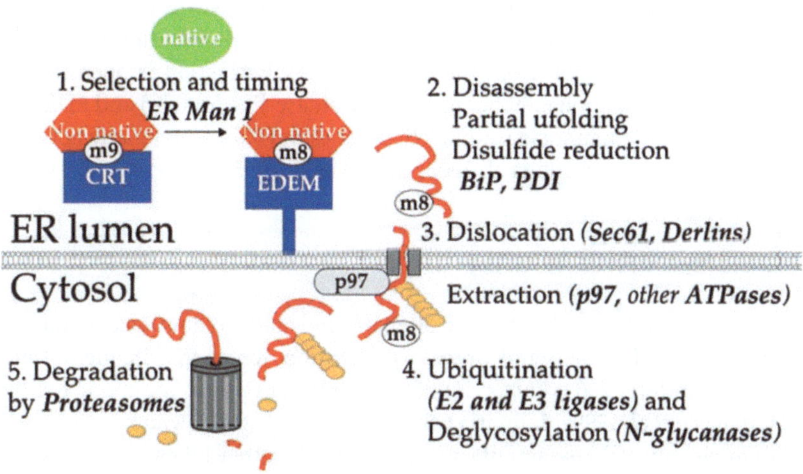

Fig. 1 The ERAD of a glycoprotein, as depicted in this figure, can be divided into five steps. First, repeated folding attempts of the substrate by the Cnx/Crt cycle are interrupted by the concerted activity of a mannose-trimming enzyme (the ER alpha-mannosidase I, ER Man I) and a mannose-binding lectin (EDEM). Second, the aberrantly folded polypeptide has to be disassembled and partially unfolded (e.g., the wrong disulfide bonds formed during the folding attempts have to be reduced) in order to (third) allow efficient dislocation across the ER membrane via Sec61, derlins and possibly other unidentified proteinaceous channels. Extraction is facilitated by cytosolic ATPases, such as p97. Fourth, E2 and E3 ubiquitin ligases often associated with the cytosolic face of the ER membrane add polyubiquitin chains to the substrate that after removal of the glycans moieties by N-glycanases are targeted to proteasomes (fifth) for degradation. Alternative, nonproteasomal pathways of degradation can also exist

2.1 Specificity in Substrate Selection

The issue of specificity in substrate selection is crucial to folding and quality control. Folding and assembly intermediates expose surfaces that elicit the attention of the ER chaperone networks. The problem becomes that of deciding when the attempts to fold a given molecule must be stopped so as to prevent accumulation. For glycoproteins, the discrimination between newly made unfolded and terminally misfolded is based on N-glycan processing. Based on the observation that inhibition of α-mannosidases stabilizes terminally misfolded glycoproteins (Su et al. 1993), the existence of a mannose timer was proposed (Helenius 1994). The model has been refined in recent years and now describes the importance of the central mannose of N-glycans (isomer B) in determining the fate of glycoproteins. Preservation of this mannose delays disposal; in contrast, accelerating its removal (for instance by up-regulation of the cleaving enzyme, ER α-mannosidase I) results in premature degradation of folding intermediates (Wu et al. 2003). Exposure of glycans with a reduced number of mannoses by non-native proteins causes interruption of folding attempts and extraction from the Cnx/Crt cycle. Extraction is operated by EDEM (ER degradation enhancing α-mannosidase-like protein) and EDEM2, stress-regulated, mannosidase-like chaperones devoid of enzymatic activity (Molinari et al. 2003; Oda et al. 2003; Olivari et al. 2005; Mast et al. 2005).

2.2 Disassembly and Partial Unfolding

Disassembly and partial unfolding often depend on the reduction of disulfide bonds in the ER lumen and precede dislocation across the ER membrane (Fagioli and Sitia 2001; Tortorella et al. 1998). These steps are mediated by PDI (Gillece et al. 1999; Molinari et al. 2002; Tsai et al. 2001), which is endowed with intrinsic unfoldase activity, and is assisted by other ER chaperones such as BiP. Misfolded polypeptides may also undergo aggregation in the ER lumen. In this case, both inter- and intramolecular disulfide bonds need to be reduced to promote cytosolic dislocation (Molinari et al., 2002).

2.3 Dislocation or Retrotranslocation

Dislocation, or retrotranslocation, has long been known to involve Sec61, a proteinaceous channel also utilized by proteins entering co-translationally into the ER. Additional proteins, termed derlins, have been recently implicated (Lilley and Ploegh 2004; Ye et al. 2004). Particularly for soluble substrates, extraction from the ER lumen involves energy. The p97 AAA-ATPase has been shown to play a role, perhaps generating force from the cytosol. In addition, many substrates accumulate in the ER lumen when proteasome activity is blocked, indicating that dislocation and degradation are coupled reactions (Mancini et al. 2000; Mayer et al. 1998; Molinari et al. 2002). The mechanisms connecting the two phases are unclear, but the ATPases associated with the proteasomes are likely to play a role, perhaps in association with p97.

2.4
N-glycanase and Ubiquitin Conjugation

Glycoprotein substrates that also dislocate in the presence of proteasome inhibitors accumulate in the cytosol in a deglycosylated and generally polyubiquitinated state. Moreover, in cells lacking a specific E3 ligase, the substrate accumulates in the ER and causes cytotoxicity (Imai et al. 2001). This implies that N-glycanase and ubiquitin conjugation activities precede targeting to the proteasomes. Evidence has been provided indicating that polyubiquitination is important for dislocation, possibly preventing backward movement of the dislocating protein into the ER lumen (de Virgilio et al. 1998). It is not clear whether N-glycanases are part of the extraction complex (Katiyar et al. 2004).

In summary, following repeated folding attempts in the Cnx/Crt cycle (a lag between synthesis and degradation) interrupted by mannose processing, EDEM diverts glycoproteins to unfolding and dislocation (Fig. 1).

Mechanistically, the sequential cleavage of mannoses from N-glycans affects retention in the Cnx/Crt cycle by decreasing the capacity of glucosidase II and GT, the enzymes regulating dissociation from and reassociation to Cnx and Crt, to act on folding-incompetent polypeptides. The role of EDEM as an acceptor of misfolded glycoproteins released from the Cnx/Crt cycle has been recently elucidated (Molinari et al. 2003; Oda et al. 2003). It has been demonstrated that the intracellular level of EDEM regulates the kinetics of dislocation by determining the permanence of misfolded proteins in the Cnx/Crt cycle. Importantly, the level of EDEM is adapted to the cargo load via unfolded protein response (UPR) signaling pathways, so as to ensure optimal disposal of the waste produced during protein synthesis and to limit stress (Yoshida et al. 2003). Whether EDEM represents the last acceptor of ERAD candidates before their dislocation into the cytosol or instead EDEM regulates the transfer of substrates from the folding into the unfolding cycles remains an open question. It is generally thought that ER mannosidase I works upstream of EDEM and actually prepares the ERAD candidate for association with EDEM by reducing the number of mannoses on N-glycans. However, the possibility exists that EDEM directly competes with Cnx and Crt for substrate binding.

In this scenario, the function of ER mannosidase I could be that of extracting ERAD candidates from EDEM to facilitate dislocation. How proteins that carry no glycans are handled remains unknown.

2.5 Cytosolic Proteasomes

While there is general agreement that the terminal point of ERAD is cytosolic proteasomes, the possibility of alternative pathways regulated by other cytosolic proteases (Glas et al. 1998) or by fusion of specialized ER subdomains with lysosomes remains open.

3 The Unfolded Protein Response in Disease and Development

Recent advances in understanding ER homeostasis show that cargo load, folding and transport efficiency, and the activity of the ERAD machinery are finely tuned at both the transcriptional and translational levels and point to a regulatory role of UPR-like pathways. Experimentally, severe ER stress can be induced by substances that inhibit ER protein folding. This can be achieved by preventing disulfide bond formation (e.g., by DTT), glycosylation (tunicamycin) or calcium homeostasis (thapsigargin). Under these harsh conditions, ER stress triggered by accumulation of abundant misfolded polypeptides activates three resident sensor molecules spanning the ER membrane: the kinase PERK, the transcription factor ATF6, and the kinase/endoribonuclease IRE1 (Harding et al. 2002; Kaufman 2002). PERK activation attenuates protein synthesis, reducing the load on the factory. ATF6 and IRE1 activate the transcription of many genes involved in promoting protein folding, degradation and transport, and other metabolic pathways (Travers et al. 2000). An elegant mechanism for timing these events has been proposed (Yoshida et al. 2003). The slower induction of Xbp1 (which, in contrast to Perk and ATF6, needs an RNA processing step catalyzed by the stress sensor IRE1) and the identification of EDEM as an Xbp1-regulated gene, led to the proposal that the Xbp1 phase of the response mainly deals with increasing degradation capacity. After a first attempt to solve the problem by lowering the load (PERK activation) and increasing folding efficiency (ATF6 activation), factors that clear the field (EDEM) are activated. In case of failure, multiple UPR-dependent pathways can cause apoptosis.

Living cells in healthy organisms probably seldom encounter such extreme conditions as those induced with global pharmacological blockade of protein folding. Certainly, successful developmental programs are executed that entail the massive production of proteins, some of which are inherently difficult to fold. Under more physiologically stressful conditions caused by increased synthesis of normal secretory proteins during organogenesis or differentiation, the folding machinery (ATF6) and the ERAD machinery (IRE1/Xbp1) can both be strengthened so as to achieve a level that prevents apoptotic programs from being activated.

Intriguing links between proteasomes, ERAD, the UPR, and apoptosis are revealed by several human diseases. In autosomal recessive juvenile parkinsonism (AR-JP), the loss of functional Parkin, a specific E3 ligase, seems to cause intracellular accumulation of PAEL receptors and UPR-dependent apoptosis of cholinergic neurons (Imai et al. 2001). In polyglutamine diseases, the inefficient cleavage between adjacent glutamines causes occlusion of proteasomes. Constipation of the ERAD machinery may lead to a chronic UPR and cell death (Ozcan et al. 2004).

UPR-like processes are key regulators of physiologic processes such as organogenesis (e.g., liver and bone development) or development of plasma cells and other professional secretors (Ma and Hendershot 2003). The IRE1-activated transcription factor Xbp1 is required for survival and/or development of secretory cells and organs (Clauss et al. 1993; Iwakoshi et al. 2003; Reimold et al. 2000, 2001).

An intriguing question that stems from appreciating the UPR's role in developing a secretory phenotype is whether and how cells can selectively activate the IRE1 and ATF6 pathways required to increase lipid and protein synthesis and to enhance metabolism without concomitantly activating the PERK pathway, which would attenuate protein synthesis and thus be detrimental for a cell devoted to massive production of a secretory protein. At present, it seems in fact that all three ER stress sensors are activated by a common mechanism relying on sequestration of the ER-resident chaperone BiP by aberrant proteins accumulating in the ER lumen (Bertolotti et al. 2000; Shen et al. 2001).

4
An Efficient and Adaptable Degradation Machinery Is Required to Maintain Folding Efficiency and Secretory Capacity

The current models describing UPR and UPR-like events in mammalian cells are focused on the involvement of these processes in preparation and adaptation of the folding/secretion machinery to the increased synthesis of ER client proteins, as during plasma cell differentiation (Gass et al. 2004; van

Anken et al. 2003). Cumulating evidence shows that the ERAD machinery is essential for efficient ER protein folding and secretion. First, it has been shown that the IRE1/Xbp1 stress pathway regulates the intraluminal level of EDEM (Eriksson et al. 2004; Yoshida et al. 2003), alpha mannosidase I (Shaffer et al. 2004), and other ERAD mediators. Second, cells depleted of Xbp1 and revealing defective regulation of the intraluminal level of EDEM progressively accumulated aberrant by-products of protein synthesis in the ER lumen. This eventually led to substantial loss of folding efficiency and strongly decreased protein secretion. Quite surprisingly, folding capacity was re-established in these cells upon EDEM transfection. These data highlight a direct link between functional ERAD machinery and capacity of cells to maintain efficient glycoprotein folding and secretion (Eriksson et al. 2004).

5
Professional Secretors

Certain cells are specialized in the production and release of secretory proteins in massive quantities. Their products vary considerably in terms of biochemical properties, function, and kinetics of secretion (Table 2). For instance, antibodies are secreted constitutively, while insulin and digestive enzymes are released in a regulated way. Despite these differences, however, a com-

Table 2 Examples of professional secretors

Cell type	Main products	Type
Plasma cells	Ig	Constitutive
Plasmocytoid DC	αIFN	Constitutive
Endocrine pancreas		
	Insulin (β cells)	Regulated synthesis and secretion
	Glucagon (α cells)	
Exocrine pancreas	Enzymes	Regulated
Salivary gland	Amylase, lysozyme, etc.	Mainly regulated
Liver	Albumin, Transferrin, and many others	Mainly constitutive (regulated production of acute phase proteins)
Fibroblasts	Collagen	Constitutive (requires hsp47 and Vit C)
Neurons	Vasopressin, oxytocin, and many others	Mainly regulated

mon trait of professional secretory cells is the remarkable development of the rough ER, where secreted proteins are made and assembled.

A single plasma cell is capable of secreting thousands of antibodies per second (de StGroth and Scheidegger 1980). On top of the remarkable metabolic and energy requirements, the cell faces a serious redox problem. Since an IgM pentamer contains about 100 disulfide bonds, 200,000 cysteines enter the ER in the reduced state and leave it oxidized every second. Ero1 molecules, which are rate limiting in oxidative folding, utilize molecular oxygen as the terminal electron acceptor, producing large quantities of H_2O_2 that need to be scavenged. In addition, in view of the intrinsic difficulty in forming polymers, a considerable fraction of secretory IgM is degraded even in plasma cells (Fagioli and Sitia 2001). The reduction of disulfide bonds that precede dislocation/degradation consumes additional glutathione, increasing the risk of oxidative stress. To comply with these requirements, plasma cell differentiation entails the up-regulation of many genes involved in energy production, protein and lipid synthesis, and redox homeostasis (van Anken et al. 2003). These, as well as ER resident proteins, are abundantly expressed in many other secretory tissues (Pagani et al. 2000). Are they under common control mechanisms? Recent evidence points to a role of Xbp1 in determining the secretory phenotype, regardless of the lineage of origin. Thus, overexpression of spliced Xbp1 (but not of the unspliced isoform) triggers a series of genetic and morphologic changes that lead to an overall increase in cell size, largely attributable to de novo synthesis of ER and other secretory organelles (Shaffer et al. 2004; Sriburi et al. 2004). The genes up-regulated by active Xbp1 largely overlap with UPR-dependent ones, but not completely. This raises the intriguing possibility that a physiologic UPR-like pathway exists that regulates development. In the case of B cell differentiation, Blimp1, a transcription factor whose overexpression is sufficient to induce a plasmacytoid phenotype, activates the expression of Xbp1. It remains to be seen how Xbp1 splicing is regulated during the physiological UPR. Brewer and co-workers showed that CHOP, a pro-apoptotic factor induced by the PERK pathway, is not expressed during the mitogen-induced differentiation of normal or neoplastic B cells (Gass et al. 2002). This result is of importance, as it highlights the possibility of independently activating the Ire1 and PERK pathways. As discussed above, the latter would be obviously detrimental for the acquisition of the secretory phenotype, as it would attenuate translation. How the physiological UPR manages to selectively turn on IRE1 remains an open question.

6 Proteasome Inhibitors and Cancer

The finding that EDEM expression is controlled by Xbp1 led to the proposal that this arm of the UPR plays a primary role in ERAD (Yoshida et al. 2003). Further analyses are required to determine if such a net distinction is correct. On the one hand, the processes of ERAD are intrinsically connected with other events on either side of the ER membrane. However, data from the clinics reinforce the view that an efficient ERAD is required for folding efficiency and even survival. Multiple myeloma cells are particularly sensitive to proteasome inhibitors, and clinical trials are giving promising results for this so far incurable disease. Why is myeloma so sensitive to this new class of drugs? Among the possible mechanisms (certain survival and homing factors are proteasome substrates; see reviews), the fact that myeloma cells are professional Ig secretors should be taken into consideration in view of the findings summarized herein. Preliminary data from our laboratories suggest a correlation between the synthesis of ERAD substrates and the sensitivity to apoptosis induced by proteasome inhibitors. It will be of interest to determine whether an imbalance between load (protein synthesis) and degradative capacity is involved in limiting the lifespan of professional secretory cells. Unraveling the intricacies of the protein factory could have profound implications in biotechnology and medicine.

References

Aridor M, Balch WE (1999) Integration of endoplasmic reticulum signaling in health and disease. Nat Med 5:745–751

Bertolotti A, Zhang Y, Hendershot LM, Harding HP, Ron D (2000) Dynamic interaction of BiP and ER stress transducers in the unfolded-protein response. Nat Cell Biol 2:326–332

Bossy-Wetzel E, Schwarzenbacher R, Lipton SA (2004) Molecular pathways to neurodegeneration. Nat Med 10 [Suppl]:S2–S9

Briquet-Laugier V, Ben-Zeev O, White A, Doolittle MH (1999) cld and lec23 are disparate mutations that affect maturation of lipoprotein lipase in the endoplasmic reticulum. J Lipid Res 40:2044–2058

Clauss IM, Gravallese EM, Darling JM, Shapiro F, Glimcher MJ, Glimcher LH (1993) In situ hybridization studies suggest a role for the basic region-leucine zipper protein hXBP-1 in exocrine gland and skeletal development during mouse embryogenesis. Dev Dyn 197:146–156

De St Groth SF, Scheidegger D (1980) Production of monoclonal antibodies: strategy and tactics. J Immunol Methods 35:1–21

De Virgilio M, Weninger H, Ivessa NE (1998) Ubiquitination is required for the retrotranslocation of a short-lived luminal endoplasmic reticulum glycoprotein to the cytosol for degradation by the proteasome. J Biol Chem 273:9734–9743

Ellgaard L, Helenius A (2003) Quality control in the endoplasmic reticulum. Nat Rev Mol Cell Biol 4:181–191

Eriksson KK, Vago R, Calanca V, Galli C, Paganetti P, Molinari M (2004) EDEM contributes to maintenance of protein folding efficiency and secretory capacity. J Biol Chem 279:44600–44605

Fagioli C, Sitia R (2001) Glycoprotein quality control in the endoplasmic reticulum. Mannose trimming by endoplasmic reticulum mannosidase I times the proteasomal degradation of unassembled immunoglobulin subunits. J Biol Chem 276:12885–12892

Ferrari DM, Soling H-D (1999) The protein disulphide-isomerase family: unraveling a string of folds. Biochem J 339:1–10

Freedman RB, Klappa P, Ruddock LW (2002) Protein disulfide isomerases exploit synergy between catalytic and specific binding domains. EMBO Rep 3:136–140

Gass JN, Gifford NM, Brewer JW (2002) Activation of an unfolded protein response during differentiation of antibody-secreting B cells. J Biol Chem 277:49047–49054

Gass JN, Gunn KE, Sriburi R, Brewer JW (2004) Stressed-out B cells? Plasma-cell differentiation and the unfolded protein response. Trends Immunol 25:17–24

Gillece P, Luz, JM, Lennarz, WJ, de La Cruz FJ, Romisch K (1999) Export of a cysteine-free misfolded secretory protein from the endoplasmic reticulum for degradation requires interaction with protein disulfide isomerase. J Cell Biol 147:1443–1456

Glas R, Bogyo M, McMaster JS, Gaczynska M, Ploegh HL (1998) A proteolytic system that compensates for loss of proteasome function. Nature 392:618–622

Hendershot LM, Sitia R (2004) Immunoglobulin assembly and secretion. In: Alt FW, Honjo T, Neuberger MS (eds) Molecular biology of B cells. Elsevier, Amsterdam, pp 261–273

Harding HP, Ron D (2002) Endoplasmic reticulum stress and the development of diabetes: a review. Diabetes 51 [Suppl 3]:S455–S461

Harding HP, Calfon M, Urano F, Novoa I, Ron D (2002) Transcriptional and translational control in the mammalian unfolded protein response. Annu Rev Cell Dev Biol 18:575–599

Helenius A (1994) How N-linked oligosaccharides affect glycoprotein folding in the endoplasmic reticulum. Mol Biol Cell 5:253–265

Helenius A, Aebi M (2004) Roles of N-linked glycans in the endoplasmic reticulum. Annu Rev Biochem 73:1019–1049

Imai Y, Soda M, Inoue H, Hattori N, Mizuno Y, Takahashi R (2001) An unfolded putative transmembrane polypeptide, which can lead to endoplasmic reticulum stress, is a substrate of Parkin. Cell 105:891–902

Iwakoshi NN, Lee AH, Vallabhajosyula P, Otipoby KL, Rajewsky K, Glimcher LH (2003) Plasma cell differentiation and the unfolded protein response intersect at the transcription factor XBP-1. Nat Immunol 4:321–329

Jansens A, van Duijn E, Braakman I (2002) Coordinated nonvectorial folding in a newly synthesized multidomain protein. Science 298:2401–2403

Katiyar S, Li G, Lennarz WJ (2004) A complex between peptide: N-glycanase and two proteasome-linked proteins suggests a mechanism for the degradation of misfolded glycoproteins. Proc Natl Acad Sci U S A 101:13774–13779

Kaufman RJ (2002) Orchestrating the unfolded protein response in health and disease. J Clin Invest 110:1389–1398

Kopito RR (1999) Biosynthesis and degradation of CFTR. Physiol Rev 79:S167–S173
Lilley BN, Ploegh HL (2004) A membrane protein required for dislocation of misfolded proteins from the ER. Nature 429:834–840
Ma Y, Hendershot LM (2003) The stressful road to antibody secretion. Nat Immunol 4:310–311
Mancini R, Fagioli C, Fra AM, Maggioni C, Sitia R (2000) Degradation of unassembled soluble Ig subunits by cytosolic proteasomes: evidence that retrotranslocation and degradation are coupled events. FASEB J 14:769–778
Mast SW, Diekman K, Karaveg K, Davis A, Sifers RN, Moremen KW (2005) Human EDEM2, a novel homolog of family 47 glycosidases, is involved in ER-associated degradation of glycoproteins. Glycobiology 15:421–436
Matsuoka Y, Kubota H, Adachi E, Nagai N, Marutani T, Hosokawa N, Nagata K (2004) Insufficient folding of type IV collagen and formation of abnormal basement membrane-like structure in embryoid bodies derived from Hsp47-null embryonic stem cells. Mol Biol Cell 15:4467–4475
Mayer TU, Braun T, Jentsch S (1998) Role of the proteasome in membrane extraction of a short-lived ER-transmembrane protein. EMBO J 17:3251–3257
Melchers F, ten Boekel E, Seidl T, Kong XC, Yamagami T, Onishi K, Shimizu T, Rolink AG, Andersson J (2000) Repertoire selection by pre-B-cell receptors and B-cell receptors, and genetic control of B-cell development from immature to mature B cells. Immunol Rev 175:33–46
Molinari M, Helenius A (2000) Chaperone selection during glycoprotein translocation into the endoplasmic reticulum. Science 288:331–333
Molinari M, Galli C, Piccaluga V, Pieren M, Paganetti P (2002) Sequential assistance of molecular chaperones and transient formation of covalent complexes during protein degradation from the ER. J Cell Biol 158:247–257
Molinari M, Calanca V, Galli C, Lucca P, Paganetti P (2003) Role of EDEM in the release of misfolded glycoproteins from the calnexin cycle. Science 299:1397–1400
Oda Y, Hosokawa N, Wada I, Nagata K (2003) EDEM as an acceptor of terminally misfolded glycoproteins released from calnexin. Science 299:1394–1397
Olivari S, Galli C, Alanen H, Ruddock L, Molinari M (2005) A novel stress-induced EDEM variant regulating endoplasmic reticulum-associated glycoprotein degradation. J Biol Chem 280:2424–2428
Ozcan U, Cao Q, Yilmaz E, Lee AH, Iwakoshi NN, Ozdelen E, Tuncman G, Gorgun C, Glimcher LH, Hotamisligil GS (2004) Endoplasmic reticulum stress links obesity, insulin action, and type 2 diabetes. Science 306:457–461
Pagani M, Fabbri M, Benedetti C, Fassio A, Pilati S, Bulleid NJ, Cabibbo A, Sitia R (2000) Endoplasmic reticulum oxidoreductin 1-lbeta (ERO1-Lbeta), a human gene induced in the course of the unfolded protein response. J Biol Chem 275:23685–23692
Perlmutter DH (1996) Alpha-1-antitrypsin deficiency: biochemistry and clinical manifestations. Ann Med 28:385–394
Reimold AM, Etkin A, Clauss I, Perkins A, Friend DS, Zhang J, Horton HF, Scott A, Orkin SH, Byrne MC et al (2000) An essential role in liver development for transcription factor XBP-1. Genes Dev 14:152–157
Reimold AM, Iwakoshi NN, Manis J, Vallabhajosyula P, Szomolanyi-Tsuda E, Gravallese EM, Friend D, Grusby MJ, Alt F, Glimcher LH (2001) Plasma cell differentiation requires the transcription factor XBP-1. Nature 412:300–307

Shaffer AL, Shapiro-Shelef M, Iwakoshi NN, Lee AH, Qian SB, Zhao H, Yu X, Yang L, Tan BK, Rosenwald A et al (2004) XBP1, downstream of Blimp-1, expands the secretory apparatus and other organelles, and increases protein synthesis in plasma cell differentiation. Immunity 21:81–93

Shang J, Korner C, Freeze H, Lehrman MA (2002) Extension of lipid-linked oligosaccharides is a high-priority aspect of the unfolded protein response: endoplasmic reticulum stress in type I congenital disorder of glycosylation fibroblasts. Glycobiology 12:307–317

Shen X, Ellis RE, Lee K, Liu CY, Yang K, Solomon A, Yoshida H, Morimoto R, Kurnit DM, Mori K, Kaufman RJ (2001) Complementary signaling pathways regulate the unfolded protein response and are required for C. elegans development. Cell 107:893–903

Sitia R, Braakman I (2003) Quality control in the endoplasmic reticulum protein factory. Nature 426:891–894

Sriburi R, Jackowski S, Mori K, Brewer JW (2004) XBP1: a link between the unfolded protein response, lipid biosynthesis, and biogenesis of the endoplasmic reticulum. J Cell Biol 167:35–41

Su K, Stoller T, Rocco J, Zemsky J, Green R (1993) Pre-Golgi degradation of yeast prepro-alpha-factor expressed in a mammalian cell. Influence of cell type-specific oligosaccharide processing on intracellular fate. J Biol Chem 268:14301–14309

Tortorella D, Story CM, Huppa JB, Wiertz EJ, Jones TR, Bacik I, Bennink JR, Yewdell JW, Ploegh HL (1998) Dislocation of type I membrane proteins from the ER to the cytosol is sensitive to changes in redox potential. J Cell Biol 142:365–376

Travers KJ, Patil CK, Wodicka L, Lockhart DJ, Weissman JS, Walter P (2000) Functional and genomic analyses reveal an essential coordination between the unfolded protein response and ER-associated degradation. Cell 101:249–258

Trombetta ES, Parodi AJ (2003) Quality control and protein folding in the secretory pathway. Annu Rev Cell Dev Biol 19:649–676

Tsai B, Rodighiero C, Lencer WI, Rapoport TA (2001) Protein disulfide isomerase acts as a redox-dependent chaperone to unfold cholera toxin. Cell 104:937–948

Van Anken E, Romijn EP, Maggioni C, Mezghrani A, Sitia R, Braakman I, Heck AJ (2003) Sequential waves of functionally related proteins are expressed when B cells prepare for antibody secretion. Immunity 18:243–253

Witzig TE, Wahner-Roedler DL (2002) Heavy chain disease. Curr Treat Options Oncol 3:247–254

Wu Y, Swulius MT, Moremen KW, Sifers RN (2003) Elucidation of the molecular logic by which misfolded alpha 1-antitrypsin is preferentially selected for degradation. Proc Natl Acad Sci U S A 100:8229–8234

Ye Y, Shibata Y, Yun C, Ron D, Rapoport TA (2004) A membrane protein complex mediates retro-translocation from the ER lumen into the cytosol. Nature 429:841–847

Yoshida H, Matsui T, Hosokawa N, Kaufman RJ, Nagata K, Mori K (2003) A time-dependent phase shift in the mammalian unfolded protein response. Dev Cell 4:265–271

CTMI (2006) 300:17–40

Recognition and Delivery of ERAD Substrates to the Proteasome and Alternative Paths for Cell Survival

A. A. McCracken[1] (✉) · J. L. Brodsky[2]

[1]Biology Department, University of Nevada, Reno, NV 89557, USA
mccracke@unr.edu

[2]Department of Biological Sciences, University of Pittsburgh,
Pittsburgh, PA 15260, USA

Abstract Endoplasmic reticulum-associated protein degradation (ERAD) is a protein quality control mechanism that minimizes the detrimental effects of protein misfolding in the secretory pathway. Molecular chaperones and ER lumenal lectins are essential components of this process because they maintain the solubility of unfolded proteins and can target ERAD substrates to the cytoplasmic proteasome. Other factors are likely required to aid in the selection of ERAD substrates, and distinct proteinaceous machineries are required for substrate retrotranslocation/dislocation from the ER and proteasome targeting. When the capacity of the ERAD machinery is exceeded or compromised, multiple degradative routes can be enlisted to prevent the detrimental consequences of ERAD substrate accumulation, which include cell death and disease.

1 Introduction

The function of almost all proteins requires that their native states are attained, but the pathway that proteins take to achieve these states is poorly understood. Even though it has been generally thought that the cellular protein folding machinery operates with high efficiency, recent data suggest instead that many proteins fold incorrectly during their biogenesis (Schubert et al. 2000; Varga et al. 2004, and references therein). Moreover, genetic mutations, translational errors, and intracellular stresses such as elevated temperature and altered pH or calcium concentration induce protein misfolding. Mis-folded proteins may self-assemble and form toxic aggregates, which in many cases give rise to specific human diseases (Lomas and Carrell 2002; Kopito and Ron 2000; Coughlan and Brodsky 2003). Fortunately, the cell has evolved machines that help misfolded proteins refold, and—if folding cannot be achieved—that destroy these potentially lethal agents.

Besides folding into their proper conformations, about one-quarter of all proteins in eukaryotes must be directed to specific intracellular compartments or to the external milieu before they can function (Cherry et al. 1997). Many of these secreted proteins are first imported, or "translocated" into the lumen of the endoplasmic reticulum (ER), a compartment that evolved to fold large numbers of chemically and physically diverse polypeptides. Because mutations in many secreted proteins lead to horrendous but common human diseases (Aridor and Hannan 2000, 2002), the mechanisms that have evolved to handle misfolded proteins in the secretion pathway have been studied intensively using genetic and biochemical attacks. Specifically, two protein quality control (QC) mechanisms are initiated in the ER to ensure that unfolded proteins are detected and that their effects are minimized. The first mechanism involves activation of the unfolded protein response (UPR), a signaling pathway that senses unfolded proteins in the ER and enhances the production of proteins that lessen ER stress and assist in protein folding (Patil and Walter 2001; Rutkowski and Kaufman 2004). The second mechanism is endoplasmic reticulum-associated protein degradation (ERAD), a pathway that delivers aberrant secreted proteins from the ER to the cytoplasm where they are degraded by the proteasome (Werner et al. 1996; McCracken et al. 1998). Therefore, ERAD also lessens the ER stress that results from misfolded protein accumulation.

Both the UPR and ERAD require a class of proteins known as molecular chaperones (reviewed in Fewell et al. 2001). Many chaperones bind to short, linear arrays of amino acids enriched for hydrophobic residues (Flynn et al. 1991; Blond-Elguindi et al. 1993; Rudiger et al. 1997, 2001). These regions often represent normally buried regions in native proteins, and if they remain solvent-exposed, protein aggregation can occur. Consequently, chaperones retain unfolded proteins in solution; however, if folding is delayed, chaperones can target these proteins for ERAD. In addition, as the concentration of unfolded proteins rises in the ER, the titration of chaperones from a UPR sensor to unfolded substrates triggers UPR induction (Bertolotti et al. 2000). Thus, ERAD and the UPR cooperate to minimize misfolded protein toxicity in the ER, and not surprisingly defects in ERAD lead to the UPR, and cells lacking key components of both ERAD and the UPR machineries are hypersensitive to further stress (Cassagrande et al. 2000; Friedlander et al. 2000; Travers et al. 2000; Ng et al. 2000).

In this review, we will first discuss the mechanisms of recognition and delivery of ERAD substrates to the proteasome. Next, we will discuss the cellular response when ERAD and the UPR are besieged by an overload of unwanted aberrant proteins, which in mammals and in yeast can result in apoptosis and apoptotic-like phenomena, respectively (Harding et al. 2000; Haynes et al. 2004). We purposely focus on recent controversies and open questions, but for more comprehensive information in this field the reader is referred to other reviews in both this volume and elsewhere (Fewell et al. 2001; Hampton 2002; Tsai et al 2002; Kostova and Wolf 2003; McCracken and Brodsky 2003).

2
Molecular Chaperones Maintain ERAD Substrate Solubility

Based on their well-known roles in protein folding, molecular chaperones emerged as the most likely mediators of ERAD substrate selection, and it was possible that prolonged chaperone interaction might redirect a protein to the ERAD pathway. A chaperone family that could provide this function is Hsp70, and the Hsp70 homolog in the ER is BiP (immunoglobulin heavy chain Binding Protein; Haas and Wabl 1983).

Hsp70 chaperones, including BiP, bind and release polypeptide substrates, an interaction that requires an ~15 kDa peptide-binding domain (PBD). C-terminal to the PBD is a poorly conserved, ~10 kDa C-terminal lid, and structural studies indicate that peptides reside in a channel formed by the PBD that is gated by the lid (Zhu et al. 1996). Hsp70s also contain an ~44 kDa N-terminal ATPase domain (Flaherty et al. 1990), and when ATP binds, the lid opens, resulting in weak peptide affinity. In contrast, when ADP is bound, the lid shuts and Hsp70s exhibit high peptide affinity (McCarty et al. 1995; Schmid et al. 1994). Peptides stimulate ATP hydrolysis, which closes the lid. Thus, ADP-ATP exchange releases the substrate.

The rate of ATP hydrolysis by Hsp70 is also enhanced significantly by Hsp40 chaperones, which can facilitate Hsp70-peptide capture (Liberek et al. 1995; Russell et al. 1999; Laufen et al. 1999). All Hsp40s contain an approximately 70-amino acid sequence called the J-domain that mediates Hsp70 binding (Gassler et al. 1998; Suh et al. 1998). Overall, whether they act alone or in combination with Hsp40s, Hsp70s maintain the solubility and thus facilitate the folding of a wide range of cytoplasmic and ER lumenal protein substrates. It seemed logical, then, that BiP might be required to prevent the aggregation of a misfolded, secreted protein, which in turn would be essential for its delivery, or retrotranslocation from the ER and into the cytoplasm.

Yeast BiP is encoded by the *KAR2* gene and binds post-translationally translocating polypeptides as they enter the ER lumen, an interaction that is essential for translocation because the chaperone is thought to "ratchet" or "pull" the polypeptide into the ER (Brodsky 1996; Rapoport et al. 1999). BiP also interacts with Sec63p, a polytopic, J-domain-containing protein, which positions BiP at the ER membrane to receive the translocating polypeptide. Interestingly, BiP and Sec63p are required for co-translational translocation (Brodsky et al. 1995; Young et al. 2001), suggesting that the BiP–Sec63p complex might regulate the translocation machinery. Using a translocation-defective *kar2* mutant, Wolf and colleagues concluded that BiP was required for ERAD because the degradation of a soluble ERAD substrate was compromised (Plemper et al. 1997). In addition, we utilized known *kar2* mutants and designed a genetic screen to isolate *kar2* alleles that are ERAD-defective but translocation-proficient (Brodsky et al. 1999; Kabani et al. 2003). Several ERAD-specific *kar2* alleles were isolated, and based on the locations of the affected residues they were predicted to alter peptide affinity. Indeed, ERAD efficiency correlated with the affinity between peptide substrates and the wild type or mutant BiPs. In addition, we found that soluble ERAD substrates aggregated in the ER of *kar2-1* and *kar2-133* mutants. As hypothesized, these data indicate that at least one of BiP's functions during ERAD in yeast is to bind and maintain aberrant polypeptides in solution. It is likely that mammalian BiP acts similarly during ERAD because it binds hydrophobic patches of misfolded proteins in the ER prior to their degradation, and the rates of BiP release from ERAD substrates and their degradation correlate (Knittler et al. 1995; Skowronek et al. 1998).

Because Hsp70 function is regulated by Hsp40s, it seemed logical that an ER resident J-domain-containing protein would also be required for ERAD, and one candidate was Sec63p. Although ERAD is modestly compromised in *sec63* mutants (Plemper et al. 1997), yeast deleted for the genes encoding two other ER lumenal Hsp40s, *SCJ1* and *JEM1*, exhibit profound ERAD defects in vitro and in vivo (Nishikawa et al. 2001). Moreover, ERAD substrates precipitate in *scj1/jem1* yeast and in microsomes prepared from the mutant strain. Because yeast deleted for *SCJ1/JEM1*were translocation-proficient, these results suggest that the BiP-Scj1p/Jem1p complex maintains ERAD substrate solubility, but that BiP interacts with a unique Hsp40 homolog (Sec63) to drive protein import. The mammalian homolog of Scj1, ERdj3, also associates with BiP and is required for secreted protein folding (Shen et al. 2002), but it is unknown if ERdj3 is involved in ERAD.

3 ERAD Substrate Selection: The Lectins

Nascent glycoproteins in the mammalian ER interact with calnexin and/or calreticulin, which are ER resident lectins that recognize the trimmed, monoglucosylated version of the oligosaccharyl side chain (Glc_1-Man_9-N-$AcGln_2$) (Ellgaard et al. 1999; Trombetta and Paroldi 2003; Kleizen and Braakman 2004). The interaction between calnexin and nascent proteins is enhanced by direct interactions with the polypeptide and with ERp57, a calnexin-associated disulfide isomerase that stabilizes immature proteins and promotes folding (Oliver et al. 1997; Frenkel et al. 2004). Consistent with a role for calnexin in ERAD in mammals are changes in ERAD efficiency when calnexin is absent or overexpressed, or when compounds are added to cells that alter the levels of Glc_1-Man_9-N-$AcGln_2$-containing glycoproteins (McCracken and Brodsky 1996; Sifers 2003).

How does calnexin facilitate ERAD? Purified, soluble fragments of calnexin retain aggregation-prone polypeptides in solution (Ihara et al. 1999), suggesting BiP-like chaperone activity, but the "calnexin cycle" and its relation to ERAD appears to be more complex. When an ER glucosidase opportunistically cleaves the terminal glucose on Glc_1-Man_9-N-$AcGln_2$, the affinity between calnexin and the glycan is lowered, and the freed protein might fold and proceed through the secretory pathway. If, however, the polypeptide cannot fold, the UDP-glucose:glycoprotein glucosyl transferase (UGGT) adds a glucose residue back to the Man_9-N-$AcGln_2$ oligosaccharide, which triggers calnexin reassociation (Helenius and Aebi 2004). Thus, UGGT redirects unfolded proteins back into the folding pathway, and may assume this role because it recognizes partially unfolded proteins and molten globules (Ritter and Helenius 2000; Caramelo et al. 2003). These data establish that the composition of the core glycan, which controls lectin/chaperone binding, regulates ER retention of misfolded proteins.

Can the cycle of calnexin association and dissociation with a misfolded protein continue indefinitely? At least for a few ERAD substrates, the answer to this question is "no". The "timer" that establishes when the calnexin cycle is broken is the opportunistic action of a mannosidase that trims Man_9 to Man_8, and overexpression of mannosidase enhances the degradation of some glycoproteins (Hosokawa et al. 2003; Wu et al. 2003). Conversion to the Man_8 glycan both reduces the affinity between the glycan and calnexin and favors interaction with a putative lectin known as ER degradation enhancing α-mannosidase-like protein (EDEM, also known as Htm1p/Mnl1p in yeast) (Braakman 2001). The interaction between EDEM and a misfolded glycoprotein precedes retrotranslocation, suggesting that EDEM links glycoprotein quality control and retrotranslocation (Oda et al. 2003; Molinari et al. 2003). However, only select ERAD substrates have been shown to utilize this pathway, and it is possible that other, as yet unknown, factors facilitate ERAD substrate selection. For example, nonglycosylated proteins do not interact efficiently with calnexin, and yeast lack UGGT. In mammals, some glycoproteins waiting for protein partners in the ER are fairly stable (Vanhove et al. 2001 and references therein), suggesting that immature glycoproteins are not degraded rapidly via ERAD; thus, EDEM selection might be overridden. Moreover, there exists a mind-boggling number of secreted protein structures, and in some cases the maturation of specific substrates is augmented by devoted chaperone-like proteins (Wang and Chang 1999; Hill and Cooper 2000). Therefore, ERAD substrate selection must exhibit great plasticity.

4
Diversity in the Selection of ERAD Substrates

If ERAD substrate selection requires more than one quality control receptor, then it is predicted that multiple routes can be taken by ERAD substrates prior to their destruction. Data supporting this premise continue to emerge. First, the degradation of different ERAD substrates in yeast exhibit unique requirements for *HRD* and *DER* gene products (Kostova and Wolf 2003; Hampton 2002). The corresponding *hrd* and *der* mutants were isolated based on their slowed degradation of two quite unique ERAD substrates, HMGCoA-R and CPY*, which are, respectively, a membrane protein whose levels are metabolically controlled and a soluble misfolded protein (Hampton et al. 1996; Knop et al. 1996). Second, ERAD diversity is apparent from differential requirements for Sec61 and components of the ubiquitination machinery (Werner et al. 1996; Walter et al. 2001; Teckman et al. 2000; Huyer et al. 2004). Third, we and others showed that BiP is dispensable for the degradation of integral membrane ERAD substrates (Plemper et al. 1998; Hill and Cooper 2000; Zhang et al. 2001; Taxis et al. 2003). Instead, cytoplasmic Hsp70-Hsp40 chaperones catalyze the ERAD of integral membrane proteins—particularly those containing large cytoplasmic polypeptide domains—but the same cytoplasmic proteins do not play a role in the retrotranslocation or degradation of soluble, lumenal proteins. Fourth, depending on the site of the misfolded domain (i.e., cytoplasm vs lumen), ERAD substrates encounter one or two check-points en route to degradation: a cytoplasmic check-point (ERAD-C) traps proteins containing aberrant cytoplasmic domains in the ER, and a subsequent lumenal check-point (ERAD-L) selects proteins that may escape from the ER and therefore have to be recycled from the *cis*-Golgi prior to degradation (Ahner and Brodsky 2004; Vashist and Ng 2004). And fifth, some misfolded secreted proteins escape the ER and travel beyond the *cis*-Golgi, but are "caught" and diverted to the lysosome/vacuole for degradation. For example, we found that wild type bovine pancreatic trypsin inhibitor (BPTI) was secreted from yeast, whereas a mutated, unstable BPTI mutant was degraded in the vacuole (Coughlan et al. 2004). It is possible that the BPTI mutant escaped ERAD because its maturation is BiP-independent, but a QC system located in the Golgi/endosome catches ERAD escapees. Interestingly, even BiP-interacting ERAD substrates can be diverted to this pathway if they are overexpressed (see below), indicating that ERAD functions in cooperation with diverse forms of ER quality control.

5
Soluble and Some Membrane Proteins May Retrotranslocate Through Sec61

When integral membrane ERAD substrates are overexpressed and/or the proteasome is inhibited in mammals, cytoplasmic "aggresomes" form, suggesting that membrane proteins, like soluble lumenal proteins, might be retrotranslocated to the cytoplasm from the membrane (Kopito 1999). How are these proteins dislocated from the ER? Studies in yeast suggest that the Sec61-containing translocation channel may be the conduit for retrotranslocation (Pilon et al. 1997; Plemper et al. 1997; Zhou and Schekman 1999). Moreover, in mammals some ERAD substrates could be found in a complex also containing Sec61 (Wiertz et al. 1996; Bebok et al. 1998). A Sec61 homolog in yeast identified as a ribosome-binding protein, Ssh1, may also facilitate retrotranslocation (Wilkinson et al. 2001). If Sec61/Ssh1 is a component of or the retrotranslocation pore, it is unknown whether subsets of Sec61 or Ssh1-containing channels are devoted either to translocation or retrotranslocation, or whether a single channel is bi-directional, perhaps integrating unique signals for translocation vs retrotranslocation.

It is also possible that Sec61 does not constitute the retrotranslocation channel, and it has been noted that some of the data implicating Sec61 as the retrotranslocon are indirect (Schekman 2004). Effects on ERAD in *sec61* or *ssh1* mutants might be explained by translocation defects or by the induction of nonspecific stresses that impact ERAD, and the formation of crosslinks to or an association with ERAD substrates might have arisen from interactions between Sec61 and translocating, but not retrotranslocating, intermediates. In support of the view that another protein may form the retrotranslocation channel, a complex containing the integral ER membrane proteins VIMP and Derlin-1 was recently found to bind to MHC-I throughout its retrotranslocation (Ye et al. 2004; Lilley and Ploegh 2004). In this system, MHC-I "dislocation" and thus degradation requires a human cytomegalovirus (H-CMV) gene product, US11, which also interacts with Derlin-1. Notably, Derlin-1 is the mammalian homolog of Der1, a factor required for the ERAD-L pathway (see above). Because Der1/Derlin-1 spans the ER membrane four times and binds p97, a cytosolic protein that drives retrotranslocation (see below), the retrotranslocation channel might be formed by Derlin-1, or it might function in concert with Sec61. However, because the degradation of many substrates is Der1-independent in yeast (Huyer et al. 2004), it remains possible that the Sec61-containing translocon is the retrotranslocon and/or that alternative models must be considered to explain membrane protein degradation.

Based on these data, the degradation of multi-spanning, integral membrane proteins can be envisaged to occur through one or a combination of three pathways (Fig. 1). If Sec61 and/or Derlin/VIMP are the retrotranslocation channel, the simplest model (Fig. 1a) is that integral membrane proteins re-enter the channel laterally and are pulled from the ER. One conceptual problem with this model is that a significant, and perhaps insurmountable energetic cost may be required to direct an integral membrane protein back into the aqueous channel, and then to extract it. It is also unknown how a membrane-spanning segment is permitted to re-enter the translocon laterally. However, the appeal of this model is that it is mechanistically translocation-in-reverse.

In another model, the proteasome clips cytoplasmic loops on misfolded integral membrane proteins and then extracts the remaining fragments directly from the membrane (Fig. 1b); conversely, the proteasome might even extract and degrade integral membrane proteins from the N- or C-terminus. The driving force for substrate extraction could be provided by chaperone-like AAA components of the proteasome and/or by a cytosolic AAA-containing complex (see below). Support for this model emerges from the observation that polypeptide loops are substrates for the mammalian proteasome (Lee et al. 2002; Liu et al. 2003), that a C-terminally anchored and polytopic membrane

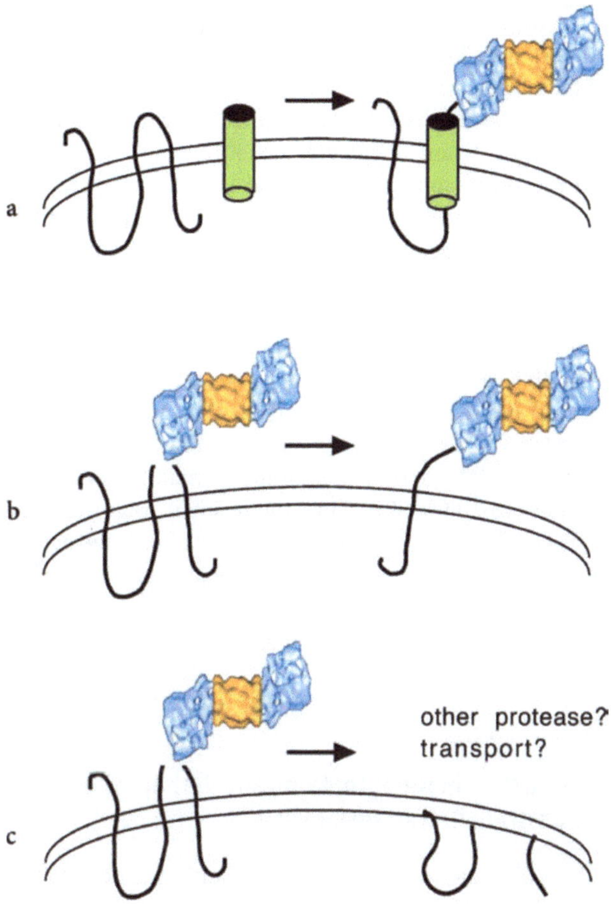

Fig. 1a–c Models for the retrotranslocation and degradation of integral membrane proteins. **a** Polytopic membrane proteins may be retrotranslocated after they re-enter the translocon (which could be comprised of Sec61, Der1/VIMP, and/or Ssh1), and export mediated by chaperones or chaperone-like proteins and/or by the proteasome. **b** The proteasome or any of the factors noted in (**a**) could provide the driving force to extract the polypeptide independent of translocon function. This process might be initiated by proteasome "clipping" (shown) or from the N- or C-terminus (not shown). **c** After proteasome cleavage and possibly "shaving" of cytoplasmic domains, the residual fragments could be degraded by another protease in the secretory pathway

protein can be degraded by the proteasome independent of Sec61 function in yeast (Walter et al. 2001; Huyer et al. 2004), and that the degradation of a misfolded membrane protein by a proteasome-like protease in bacteria has been reconstituted in the absence of a channel (Akiyama and Ito 2003).

Finally, the degradation of integral membrane ERAD substrates might require both the proteasome, which would shave cytoplasmic domains of membrane proteins, and other proteases in the secretory pathway (Fig. 1c). For example, while the proteasome might dispose of cytoplasmic loops, perhaps proteases in the ER or elsewhere in the secretory pathway mop up residual membrane fragments and luminal domains. Candidates for these proteases include the site 1 and site 2 proteases and the signal peptide peptidase (SPP) in the mammalian *cis*-Golgi and ER, respectively, each of which degrades membrane-spanning segments either within or at the membrane (Weihofen and Martoglio 2003). In support of this hypothesis, SPP was recently shown to interact with a misfolded, transmembrane domain (Crawshaw et al. 2004), but we note that yeast lack SPP. Thus, the residual fragments might traffic beyond the *cis*-Golgi and become degraded in the lysosome/vacuole. It is also possible that ill-characterized proteases act before or concurrent with the proteasome. In favor of this scenario, in few published studies do proteasome mutants or proteasome inhibitors completely block the degradation of an ERAD substrate.

6
ER-Associated Factors in the Cytoplasm, and Perhaps the Proteasome, Drive ERAD Substrate Retrotranslocation

How are ERAD substrates driven from the ER and into the cytoplasm? Three possibilities are that ERAD substrates are driven from the ER by:

1. Cytoplasmic polyubiquitinylation, which might ratchet the substrate from the ER
2. A chaperone or chaperone-like protein, viz. the action of BiP-Sec63 during post-translational translocation (see above)
3. The proteasome itself

The concept that polyubiquitinylation drives ERAD substrate retrotranslocation emerged from data indicating that extraction was halted in yeast and mammalian cells containing defects in the ubiquitin conjugation machinery, and when dominant negative forms of ubiquitin were introduced (de Virgilio et al. 1998; Shamu et al. 2001; Yu and Kopito 1999). In addition, polyubiquitin chains of sufficient lengths are required to support ERAD substrate retrotranslocation (Jarosch et al. 2002). A requirement for highly polyubiquitinated species might be necessary to prevent the reimport of ERAD substrates into the ER after they have emerged in the cytoplasm. However, at least two characterized ERAD substrates are not ubiquitin modified prior to degradation (Werner et al. 1996; Teckman et al. 2000), and the retrotranslocation of bacterial toxins that masquerade as ERAD substrates may be ubiquitin-independent (Rodighiero et al. 2002; Deeks et al. 2002). In these and perhaps other cases, retrotranslocation might be driven by chaperone and/or proteasome interaction (see below). In addition, polyubiquitin chains mediate the high-affinity binding between several ERAD substrates and a cytoplasmic multi-protein complex containing Cdc48 (also known as p97 or Valosin-containing protein, VCP, in mammals), Ufd1, and Npl4 (Tsai et al. 2002; Bays and Hampton 2002). The key component of this complex is the AAA protein Cdc48, which is a hexameric ATPase and possesses chaperone-like properties. Cdc48/p97 facilitates the dis-aggregation of ER-tethered transcription factors and maintains the solubility of unfolded, aggregation-prone proteins in solution (Braun et al. 2002; Thoms 2002). Cdc48/p97 also associates with the 19S component of the proteasome and binds to the ER membrane via Derlin-1/VIMP (Verma et al. 2000; Ye et al. 2004). Cdc48 binds first to the polypeptide backbone on ERAD substrates as they emerge from the ER, and then after a substrate becomes modified with polyubiquitin in the cytoplasm, the Cdc48-Ufd1-Npl4 complex binds more tightly via the ubiquitin moiety (Ye et al. 2003).

Do all substrates require Cdc48-Ufd1-Npl4 for retrotranslocation? As noted above, some polypeptides are delivered from the ER without polyubiquitinylation, and an alternate or complementary mediator of ERAD substrate extraction for even polyubiuqitinated species might be the proteasome itself. The 26S proteasome is composed of a 20S core particle that harbors three unique proteolytic activities, and a 19S "cap" (also known as PA700 in mammals) that binds and delivers polypeptide substrates to the core (Voges et al. 1999). The base of the cap is comprised of six AAA proteins whose ATPase activity is thought to drive polypeptide movement, as proposed for Cdc48, and regulate the opening of an aperture that limits polypeptide entry into the core (Finley et al. 2000). The cap has also been shown to exhibit chaperone-like activity; i.e., purified 19S particles retain misfolded proteins in solution (Braun et al. 1999; Strickland et al. 2000). Moreover, some studies indicated that inhibition of proteasome activity prevents substrate retrotranslocation, suggesting that retrotranslocation and degradation are tightly coupled, and that inhibition of proteasome activity slows membrane extraction and degradation of model membrane proteins, suggesting that the proteasome itself drives extraction (Mayer et al. 1998).

To explore directly whether the proteasome was necessary and sufficient to both dislocate and degrade an ERAD substrate, we incubated purified proteasomes with yeast ER-derived microsomes that had been preloaded with a soluble ERAD substrate and observed ATP-dependent retrotranslocation and degradation (Lee et al. 2004). We were also able to uncouple the retrotranslocation and degradation steps by incubating the ERAD substrate-loaded microsomes with purified 19S particles, which extracted and bound the substrate in an ATP-dependent manner, and then by adding the core 20S particle, which resulted in substrate degradation. These results indicate that the ATPase activity of the 19S particle exerts directional force to ratchet or drive a polypeptide from the ER, and can then deliver the polypeptide to the 20S particle for degradation. Clearly, future efforts will need to be devoted to recapitulate the polyubiquitinylation, degradation, and retrotranslocation of more complex ERAD substrates.

7 ERAD Substrate Overload: An Overflow Pathway that Functions Coordinately with ERAD

Because of the potential lethality that results when aberrant proteins accumulate in the secretory pathway, ER quality control is extremely efficient, yet it is not faultless and cells are at risk when protein misfolding increases. It is not surprising then that the cell possesses alternative trafficking schemes to remove unwanted proteins. In fact, some aberrant soluble proteins appear to evade ERAD and are instead conveyed, via vesicle transport, to the Golgi with subsequent sorting to the vacuole for degradation (see above, and Hong et al. 1996; Holkeri and Makarow 1998; Jorgensen et al. 1999; Arvan et al. 2002). Furthermore, the ERAD pathway may be saturated by overexpression of aberrant proteins and upon ER stress the excess protein can be degraded in the vacuole after transiting through the Golgi (Spear and Ng 2003).

Although little is known about the mechanisms and regulation of Golgi QC, it is becoming increasingly clear that at least two pathways to the vacuole can be utilized. For example, the Z variant of the human protein alpha-1 proteinase inhibitor (A1PiZ; also known as alpha-1 antitrypsin-Z) is the most common cause of juvenile liver disease in the United States and is an ERAD substrate in both yeast and mammals (Qu et al. 1996; Werner et al. 1996). However, when A1PiZ was overexpressed in yeast, ERAD was saturated and the protein was transported to the vacuole by two distinct pathways: the CPY-to-vacuole route and autophagy (Kruse et al. 2005). These findings suggest that overexpressed A1PiZ and the accompanying ER stress are sufficient to target A1PiZ to overflow pathways, and we proposed that the removal of this toxic protein by either pathway improves yeast viability.

Why are two distinct routes employed to rid the secretory pathway of excess A1PiZ? First, excess A1PiZ may be packaged into vesicles destined for the Golgi, but when the protein reaches the *trans*-Golgi it is selected by the ill-defined Golgi/endosomal QC system for transport to the vacuole. This premise is supported by the observation that misfolded A1PiZ was secreted when the CPY-to-vacuole route was blocked (by deletion of *VPS10*, *VPS30*, or *VPS38*) (Kruse et al. 2005).

The second pathway, autophagy, has been described as the major catabolic mechanism for the degradation of long-lived proteins and organelles, in contrast to the proteasome that is known to degrade short-lived and aberrant proteins of the cytoplasm, nucleus, and ER (reviewed in Pickart 2004). Autophagy is an evolutionarily conserved process implicated in a diverse number of cellular and biological phenomena, and is activated in response to both extracellular and intracellular stress conditions and during starvation, cellular and tissue remodeling, and at specific developmental stages (Levine and Klionsky 2004). Because autophagy removes aged organelles, it is essential for cell survival. However, accumulating evidence supports a role for the autophagic pathway in the removal of misfolded proteins, which under certain conditions may also be essential. Indeed, autophagy can be activated to engulf aggresomes—deposits of aggregated cytoplasmic proteins—and deliver them to the vacuole for degradation (Kopito 2000; Ravikumar et al. 2002; Fortun et al. 2003).

Although autophagy is clearly required for the turnover of aggregation-prone, disease-causing cytoplasmic proteins such as huntingtin and α-synuclein (Ravikumar et al. 2002; Webb et al. 2003), a role for autophagy in the removal of aggregation-prone, disease-causing ER proteins has more recently been proposed and is supported by several observations:

1. The turnover of A1PiZ is impaired by inhibitors of autophagy.
2. There is an increase in the number of autophagosomes in human liver cells and cultured human fibroblasts of individuals expressing A1PiZ, and in liver cells from transgenic mice engineered to express A1PiZ
3. Autophagy is induced by the accumulation of A1PiZ in the ER (Teckman and Perlmutter 2000; Teckman et al. 2001, 2002). Indeed, it is well documented that ER-accumulated A1PiZ is aggregation-prone (Lomas et al. 1992; Dafforn et al. 1999), and increased levels of aggregated A1PiZ were seen in autophagy-deficient yeast strains (Kruse et al. 2005), suggesting that autophagy-mediated clearance of ER-aggregated A1PiZ is evolutionarily conserved.

Much evidence demonstrates that autophagy is a nonselective process that enwraps cytoplasm nonspecifically into a double membrane-bounded vesicle, the autophagosome, yet little is known about the membrane origin of this structure. Even though data are lacking that autophagic vesicles bud off from preexisting organelles, the prevalent hypothesis is that—rather than de novo vesicle synthesis—the ER provides the membrane lipids for autophagic vesicle formation (Levine and Klionsky 2004). Our findings and those of Perlmutter and colleagues (described above), suggesting vacuole delivery of ER-accumulated A1PiZ via autophagy, also implicates the ER as the origin of the autophagosome membrane. Thus, our working model of A1PiZ degradation by autophagy is that A1PiZ aggregates concentrate in the ER, and are captured into double membrane-bounded autophagosomes (Fig. 2). Alternatively, A1PiZ may be retrotranslocated to the cytosol where it aggregates and forms aggresomes that can be removed by autophagy. However, we do not favor this hypothesis because the presence of A1PiZ in human hepatocytes can lead to the production of membrane-bound inclusion bodies but not cytoplasmic aggresomes (Teckman and Perlmutter 2000).

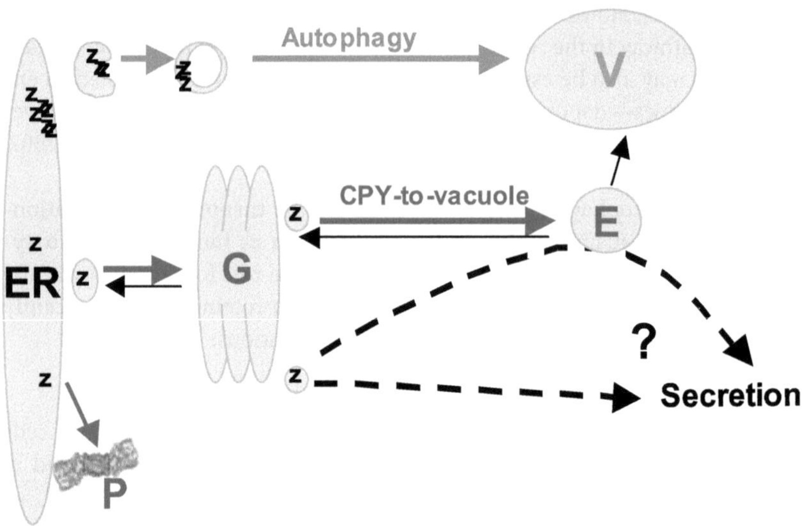

Fig. 2. Proposed model of A1PiZ quality control. At low expression levels, A1PiZ is targeted to ERAD and thus exits the ER by retrotranslocation with subsequent degradation by the proteasome (*P*). When overexpressed, excess soluble A1PiZ (*Z*) exits the ER by vesicle transport, passages the Golgi (*G*) and is sorted into the CPY-to-vacuole pathway via the endosome (*E*). If the CPY-to-vacuole pathway is blocked, soluble A1PiZ is secreted. The excess A1PiZ that aggregates (*ZZZZ*) within the ER is sent to the vacuole (*V*) via autophagy

8
Autophagy, ER Stress, and Disease

The physiological importance of autophagy is also clearly demonstrated in Danon disease, where the absence of the lysosomal membrane protein LAMP2 leads to massive accumulation of autophagosomes due to LAMP2-defective fusion of autophagosomes with the lysosome (Nishino et al. 2000). The accumulation of autophagosomes in numerous tissues, including liver, muscle, and heart, is a diagnostic feature of this disease, which is characterized by fetal cardiomyopathy and mental retardation. In view of the role of protein aggregates in disease and the possible protective function of autophagy in removal of ER aggregates, a more complete understanding of the mechanisms and regulation of autophagy is crucial.

Because ER stress can be induced by the presence of unfolded proteins and may lead to the activation of the UPR and/or ERAD, the induction of overflow quality control pathways to the vacuole, and even to cell death (Rutkowski and Kaufman 2004; Haynes et al. 2004), it follows that ER stress may be directly involved in disease pathogenesis. Common among Huntington's, Parkinson's, Alzheimer's, and ER storage diseases is the link to ER stress and the formation and accumulation of aggregated proteins in the ER (reviewed in Kim and Arvan 1998; Dimcheff et al. 2003), yet the precise role of the misfolded proteins in the pathogenesis of disease is unclear, raising important questions:

How Can Aggregates Be Explained and Prevented? A1PiZ overexpression saturates ERAD and activates overflow pathways that contribute to aggregate formation. Why some soluble misfolded proteins aggregate is not always known, although it is likely an inherent characteristic of the protein (Lomas et al. 1992). However, it is clear that the formation of both aggresomes and ER aggregates rise when the proteasome is compromised (Kopito 2000), suggesting that a change in the concentration of misfolded proteins contributes to aggregate formation. Thus, it has been noted that therapies that decrease proteasome function in vivo might have severe, secondary consequences (Ma et al. 2002).

Can a Protein's Conformation or ERAD Be Modulated to Prevent Disease? Many studies indicate that "chemical chaperones" facilitate the maturation and folding of mutated or unstable proteins (Tamarappoo and Verkman 1998; Song and Chuang 2001; Sawkar et al. 2002; Noorwez et al. 2003). Chemical chaperones are most commonly osmolytes that facilitate protein folding, either by ordering water and strengthening intramolecular bonds, or by helping solvate proteins in the process of folding. Other chemical chaperones are enzyme substrates that serve as a platform on which a protein can fold. For example, the ΔF508, a disease-causing mutant form of the cystic fibrosis transmembrane conductance regulator protein (CFTR), is not transported to the cell surface but rather is retrotranslocated from the ER and degraded by the proteasome (Gelman and Kopito 2003). However, when cells were treated with lower temperatures or chemical chaperones the mutant protein folded into an active conformation and was transported to the cell surface (Sato et al. 1996; Brown et al. 1997; Kopito 1999; Howard et al. 2003). In cell and animal studies, Hsp70 modulators have been shown to ameliorate the ΔF508 CFTR folding defect, either by modulation of ERAD or through direct effects on the folding pathway (Brodsky 2001). Some of these compounds are now in clinical trials (Zeitlin 2003).

Can the Toxic Load of Misfolded Proteins Be Reduced Through Up-Regulation of Molecular Chaperones that Disassemble Protein Aggregates? Although a mammalian "disaggregase" has not been characterized, yeast and eubacterial Hsp104 and ClpB, respectively, in cooperation with mammalian and bacterial Hsp70 homologs dissolve protein aggregates and remodel substrate proteins to their native state; other bacterial chaperones (ClpCP/ClpAP) solubilize protein aggregates for subsequent hydrolysis (Weibezahn et al. 2004). Indeed, when yeast Hsp104 was expressed in mammalian cells, heat shock-induced loss of cell viability decreased and the chaperone functioned coordinately with the endogenous Hsp70 to refold aggregated luciferase (Mosser et al. 2004).

Finally, because autophagy removes aggresomes and ER aggregates (see above), it is clear that the autophagic pathway is an essential component of protein quality control and thus normal cellular function. Hence, a more complete understanding of the mechanisms, induction, and regulation of autophagy may have direct application for therapeutic intervention of protein aggregation and other "conformational" diseases (Lomas and Carrell 2002).

References

Ahner A, Brodsky JL (2004) Checkpoints in ER-associated degradation: excuse me, which way to the proteasome? Trends Cell Biol. 14:474–478

Akiyama Y, Ito K (2003) Reconstitution of membrane proteolysis by FtsH. J Biol Chem 278:18146–18153

Aridor M, Hannan LA (2000) Traffic jam: a compendium of human diseases that affect intracellular transport processes. Traffic 1:836–851

Aridor M, Hannan LA (2002) Traffic jams II: an update of diseases of intracellular transport. Traffic 3:781–790

Arvan P, Zhao X, Ramos-Castaneda J, Chang A (2002) Secretory pathway quality control operating in Golgi, plasmalemmal, and endosomal systems. Traffic 3:771–780

Bays NW, Hampton RY (2002) Cdc48-Ufd1-Npl4: stuck in the middle with Ub. Curr Biol 12:R366–R371

Bebok Z, Mazzochi C, King SA, Hong JS, Sorscher EJ (1998) The mechanism underlying cystic fibrosis transmembrane conductance regulator transport from the endoplasmic reticulum to the proteasome includes Sec61beta and a cytosolic, deglycosylated intermediary. J Biol Chem 273:29873–29878

Bertolotti A, Zhang Y, Hendershot LM, Harding HP, Ron D (2000) Dynamic interaction of BiP and ER stress transducers in the unfolded-protein response. Nat Cell Biol 2:326–332

Blond-Elguindi S, Cwirla SE, Dower WJ, Lipshutz RJ, Sprang SR, Sambrook JF, Gething MJ (1993) Affinity panning of a library of peptides displayed on bacteriophages reveals the binding specificity of BiP. Cell 75:717–728

Braakman I (2001) A novel lectin in the secretory pathway. An elegant mechanism for glycoprotein elimination. EMBO Rep 2:666–668

Braun BC, Glickman M, Kraft R, Dahlmann B, Kloetzel PM, Finley D, Schmidt M (1999) The base of the proteasome regulatory particle exhibits chaperone-like activity. Nat Cell Biol 1:221–226

Braun S, Matuschewski K, Rape M, Thoms S, Jentsch S (2002) Role of the ubiquitin-selective CDC48(UFD1/NPL4)chaperone (segregase) in ERAD of OLE1 and other substrates. EMBO J 21:615–621

Brodsky JL (1996) Post-translational protein translocation: not all hsc70s are created equal. Trends Biochem Sci 21:122–126

Brodsky JL (2001) Chaperoning the maturation of the cystic fibrosis transmembrane conductance regulator. Am J Physiol Lung Cell Mol Physiol 281:L39–L42

Brodsky JL, Goeckeler J, Schekman R (1995) BiP and Sec63p are required for both co and posttranslational protein translocation into the yeast endoplasmic reticulum. Proc Natl Acad Sci U S A 1995 92:9643–9646

Brodsky JL, Werner ED, Dubas ME, Goeckeler JL, Kruse KB, McCracken AA (1999) The requirement for molecular chaperones during endoplasmic reticulum-associated protein degradation demonstrates that protein export and import are mechanistically distinct. J Biol Chem 274:3453–3460

Brown CR, Hong-Brown LQ, Welch WJ (1997) Strategies for correcting the delta F508 CFTR protein-folding defect. J Bioenerg Biomembr 29:491–502

Caramelo JJ, Castro OA, Alonso LG, De Prat-Gay G, Parodi AJ (2003) UDP-Glc:glycoprotein glucosyltransferase recognizes structured and solvent accessible hydrophobic patches in molten globule-like folding intermediates. Proc Natl Acad Sci U S A 100:86–91

Casagrande R, Stern P, Diehn M, Shamu C, Osario M, Zuniga M, Brown PO, Ploegh H (2000) Degradation of proteins from the ER of S. cerevisiae requires an intact unfolded protein response pathway. Mol Cell 5: 729–735

Cherry JM, Ball C, Weng S, Juvik G, Schmidt R, Adler C, Dunn B, Dwight S, Riles L, Mortimer RK, Botstein D (1997) Genetic and physical maps of Saccharomyces cerevisiae. Nature 387(6632 Suppl):67–73

Coughlan CM, Brodsky JL (2003) Yeast as a model system to investigate protein conformational diseases. Methods Mol Biol 232:77–90

Coughlan CM, Walker JL, Cochran JC, Wittrup KD, Brodsky JL (2004) Degradation of mutated bovine pancreatic trypsin inhibitor (BPTI) in the yeast vacuole suggests post-endoplasmic reticulum protein quality control. J Biol.Chem 279:15289–15297

Crawshaw SG, Martoglio B, Meacock SL, High S (2004) A misassembled transmembrane domain of a polytopic protein associates with signal peptide peptidase. Biochem J 384:9–17

Dafforn TR, Mahadeva R, Elliott PR, Sivasothy P, Lomas DA (1999) A kinetic mechanism for the polymerization of alpha1-antitrypsin. J Biol Chem 274:9548–9555

De Virgilio M, Weninger H, Ivessa NE (1998) Ubiquitination is required for the retrotranslocation of a short-lived luminal endoplasmic reticulum glycoprotein to the cytosol for degradation by the proteasome. J Biol Chem 273:9734–9743

Deeks ED, Cook JP, Day PJ, Smith DC, Roberts LM, Lord JM (2002) The low lysine content of ricin A chain reduces the risk of proteolytic degradation after translocation from the endoplasmic reticulum to the cytosol. Biochemistry 41:3405–3413

Dimcheff DE, Portis JL, Caughey B (2003) Prion proteins meet protein quality control. Trends Cell Biol 13:337–340

Ellgaard L, Molinari M, Helenius A (1999) Setting the standards: quality control in the secretory pathway. Science 286:1882–1888

Fewell SW, Travers KJ, Weissman JS, Brodsky JL (2001) The action of molecular chaperones in the early secretory pathway. Annu Rev Genet 35:149–191

Finley D, Groll M, Bajorek M, Kohler A, Moroder L, Rubin DM, Huber R, Glickman MH (2000) A gated channel into the proteasome core particle. Nat Struct Biol 7:1062–1067

Flaherty KM, DeLuca-Flaherty C, McKay DB (1990) Three-dimensional structure of the ATPase fragment of a 70 K heat-shock cognate protein. Nature 346:623–628

Flynn GC, Pohl J, Flocco MT, Rothman JE (1991) Peptide-binding specificity of the molecular chaperone BiP. Nature 353:726–730

Fortun J, Dunn WA Jr, Joy S, Li J, Notterpek L (2003) Emerging role for autophagy in the removal of aggresomes in Schwann cells. J Neurosci 23:10672–10680

Frenkel Z, Shenkman M, Kondratyev M, Lederkremer GZ (2004) Separate roles and different routing of calnexin and ERp57 in endoplasmic reticulum quality control revealed by interactions with asialoglycoprotein receptor chains. Mol Biol Cell 15:2133–142

Friedlander R, Jarosch E, Urban J, Volkwein C, Sommer T (2000) A regulatory link between ER-associated protein degradation and the unfolded-protein response. Nat Cell Biol 2:379–384

Gassler CS, Buchberger A, Laufen T, Mayer MP, Schroder H, Valencia A, Bukau B (1998) Mutations in the DnaK chaperone affecting interaction with the DnaJ cochaperone. Proc Natl Acad Sci U S A 1998 95:15229–15234

Gelman MS, Kopito RR (2003) Cystic fibrosis: premature degradation of mutant proteins as a molecular disease mechanism. Methods Mol Biol 232:27–37

Groll M, Bajorek M, Kohler A, Moroder L, Rubin DM, Huber R, Glickman MH, Finley D (2000) A gated channel into the proteasome core particle. Nat Struct Biol 7:1062–1067

Haas IG, Wabl M (1983) Immunoglobulin heavy chain binding protein. Nature 306:387–389

Hampton RY (2002) ER-associated degradation in protein quality control and cellular regulation. Curr Opin Cell Biol 14:476–482

Hampton RY, Gardner RG, Rine J (1996) Role of 26S proteasome and HRD genes in the degradation of 3-hydroxy-3-methylglutaryl-CoA reductase, an integral endoplasmic reticulum membrane protein. Mol Biol Cell 7:2029–2044

Harding HP, Novoa I, Zhang Y, Zeng H, Wek R, Schapira M, Ron D (2000) Regulated translation initiation controls stress-induced gene expression in mammalian cells. Mol Cell 6:1099–1108

Haynes CM, Titus EA, Cooper AA (2004) Degradation of misfolded proteins prevents ER-derived oxidative stress and cell death. Mol Cell 15:767–776

Helenius A, Aebi M (2004) Roles of N-linked glycans in the endoplasmic reticulum. Annu Rev Biochem 73:1019–1049

Hill K, Cooper AA (2000) Degradation of unassembled Vph1p reveals novel aspects of the yeast ER quality control system. EMBO J 19:550–561

Holkeri H, Makarow M (1998) Different degradation pathways for heterologous glycoproteins in yeast. FEBS Lett 429:162–166

Hong E, Davidson AR, Kaiser CA (1996) A pathway for targeting soluble misfolded proteins to the yeast vacuole. J Cell Biol 135:623–633

Hosokawa N, Tremblay LO, You Z, Herscovics A, Wada I, Nagata K (2003) Enhancement of endoplasmic reticulum (ER) degradation of misfolded Null Hong Kong alpha1-antitrypsin by human ER mannosidase I. J Biol Chem 278:26287–26294

Howard M, Fischer H, Roux J, Santos BC, Gullans SR, Yancey PH, Welch WJ (2003) Mammalian osmolytes and S-nitrosoglutathione promote Delta F508 cystic fibrosis transmembrane conductance regulator (CFTR) protein maturation and function. J Biol Chem 278:35159–35167

Huyer G, Piluek WF, Fansler Z, Kreft SG, Hochstrasser M, Brodsky JL, Michaelis S (2004) Distinct machinery is required in Saccharomyces cerevisiae for the endoplasmic reticulum-associated degradation of a multispanning membrane protein and a soluble luminal protein. J Biol Chem 279:38369–38378

Ihara Y, Cohen-Doyle MF, Saito Y, Williams DB (1999) Calnexin discriminates between protein conformational states and functions as a molecular chaperone in vitro. Mol Cell 4:331–341

Jarosch E, Geiss-Friedlander R, Meusser B, Walter J, Sommer T (2002) Protein dislocation from the endoplasmic reticulum—pulling out the suspect. Traffic 3:530–536

Jorgensen MU, Emr SD, Winther JR (1999) Ligand recognition and domain structure of Vps10p, a vacuolar protein sorting receptor in Saccharomyces cerevisiae. Eur J Biochem 260:461–469

Kabani M, Kelley SS, Morrow MW, Montgomery DL, Sivendran R, Rose MD, Gierasch LM, Brodsky JL (2003) Dependence of endoplasmic reticulum-associated degradation on the peptide binding domain and concentration of BiP. Mol Biol Cell 14:3437–3448

Kim PS, Arvan P (1998) Endocrinopathies in the family of endoplasmic reticulum (ER) storage diseases: disorders of protein trafficking and the role of ER molecular chaperones. Endocr Rev 19:173–202

Kleizen B, Braakman I (2004) Protein folding and quality control in the endoplasmic reticulum. Curr Opin Cell Biol 16:343–349

Knittler MR, Dirks S, Haas IG (1995) Molecular chaperones involved in protein degradation in the endoplasmic reticulum: quantitative interaction of the heat shock cognate protein BiP with partially folded immunoglobulin light chains that are degraded in the endoplasmic reticulum. Proc Natl Acad Sci U S A 92:1764–1768

Knop M, Hauser N, Wolf DH (1996) N-Glycosylation affects endoplasmic reticulum degradation of a mutated derivative of carboxypeptidase yscY in yeast. Yeast 12:1229–1238

Kopito RR (1999) Biosynthesis and degradation of CFTR. Physiol Rev 79 [1 Suppl]:S167–S173

Kopito RR (2000) Aggresomes, inclusion bodies and protein aggregation. Trends Cell Biol 10:524–530

Kopito RR, Ron D (2000) Conformational disease. Nat Cell Biol 2:E207–E209

Kostova Z, Wolf DH (2003) For whom the bell tolls: protein quality control of the endoplasmic reticulum and the ubiquitin-proteasome connection. EMBO J 22:2309–2317

Kruse KB, Brodsky JB, McCracken AA (2005) Characterization of an ERAD gene as *VPS30/ATG6* reveals two alternative and functionally distinct protein quality control pathways: one for soluble A1PiZ and another for aggregates of A1PiZ. Mol Biol Cell (in press)

Laufen T, Mayer MP, Beisel C, Klostermeier D, Mogk A, Reinstein J, Bukau B (1999) Mechanism of regulation of hsp70 chaperones by DnaJ cochaperones. Proc Natl Acad Sci U S A 96:5452–5457

Lee C, Prakash S, Matouschek A (2002) Concurrent translocation of multiple polypeptide chains through the proteasomal degradation channel. J Biol Chem 277:34760–34765

Lee RJ, Liu CW, Harty C, McCracken AA, Latterich M, Romisch K, DeMartino GN, Thomas PJ, Brodsky JL (2004) Uncoupling retro-translocation and degradation in the ER-associated degradation of a soluble protein. EMBO J 23:2206–2215

Levine B, Klionsky DJ (2004) Development by self-digestion: molecular mechanisms and biological functions of autophagy. Dev Cell 6:463–477

Liberek K, Wall D, Georgopoulos C (1995) The DnaJ chaperone catalytically activates the DnaK chaperone to preferentially bind the sigma 32 heat shock transcriptional regulator. Proc Natl Acad Sci U S A 92:6224–6228

Lilley BN, Ploegh HL (2004) A membrane protein required for dislocation of misfolded proteins from the ER. Nature 429:834–840

Liu CW, Corboy MJ, DeMartino GN, Thomas PJ (2003) Endoproteolytic activity of the proteasome. Science 299:408–411

Lomas DA, Carrell RW (2002) Serpinopathies and the conformational dementias. Nat Rev Genet 3:759–768

Lomas DA, Evans DL, Finch JT, Carrell RW (1992) The mechanism of Z alpha 1-antitrypsin accumulation in the liver. Nature 357:605–607

Ma J, Wollmann R, Lindquist S (2002) Neurotoxicity and neurodegeneration when PrP accumulates in the cytosol. Science 298:1781–1785

Mayer TU, Braun T, Jentsch S (1998) Role of the proteasome in membrane extraction of a short-lived ER-transmembrane protein. EMBO J 17:3251–3257

McCarty JS, Buchberger A, Reinstein J, Bukau B (1995) The role of ATP in the functional cycle of the DnaK chaperone system. J Mol Biol 249:126–137

McCracken AA, Brodsky JL (1996) Assembly of ER-associated protein degradation in vitro: dependence on cytosol, calnexin, and ATP. J Cell Biol 132:291–298
McCracken AA, Brodsky JL (2003) Evolving questions and paradigm shifts in endoplasmic-reticulum-associated degradation (ERAD). Bioessays 25:868–877
McCracken AA, Werner ED, Brodsky JL (1998) Endoplasmic reticulum-associated protein degradation: an unconventional route to a familiar fate. Adv Mol Cell Biol 27:167–200
Molinari M, Calanca V, Galli C, Lucca P, Paganetti P (2003) Role of EDEM in the release of misfolded glycoproteins from the calnexin cycle. Science 299:1397–1400
Mosser DD, Ho S, Glover JR (2004) Saccharomyces cerevisiae Hsp104 enhances the chaperone capacity of human cells and inhibits heat stress-induced proapoptotic signaling. Biochemistry 43:8107–8115
Ng DT, Spear ED, Walter P (2000) The unfolded protein response regulates multiple aspects of secretory and membrane protein biogenesis and endoplasmic reticulum quality control. J Cell Biol 150:77–88
Nishikawa SI, Fewell SW, Kato Y, Brodsky JL, Endo T (2001) Molecular chaperones in the yeast endoplasmic reticulum maintain the solubility of proteins for retro-translocation and degradation. J Cell Biol 153:1061–1070
Nishino I, Fu J, Tanji K, Yamada T, Shimojo S, Koori T, Mora M, Riggs JE, Oh SJ, Koga Y, Sue CM, Yamamoto A, Murakami N, Shanske S, Byrne E, Bonilla E, Nonaka I, DiMauro S, Hirano M (2000) Primary LAMP-2 deficiency causes X-linked vacuolar cardiomyopathy and myopathy (Danon disease). Nature 406:906–910
Noorwez SM, Kuksa V, Imanishi Y, Zhu L, Filipek S, Palczewski K, Kaushal S (2003) Pharmacological chaperone-mediated in vivo folding and stabilization of the P23H-opsin mutant associated with autosomal dominant retinitis pigmentosa. J Biol Chem 278:14442–14450
Oda Y, Hosokawa N, Wada I, Nagata K (2003) EDEM as an acceptor of terminally misfolded glycoproteins released from calnexin. Science 299:1394–1397
Oliver JD, van der Wal FJ, Bulleid NJ, High S (1997) Interaction of the thiol-dependent reductase ERp57 with nascent glycoproteins. Science 275:86–88
Patil C, Walter P (2001) Intracellular signaling from the endoplasmic reticulum to the nucleus: the unfolded protein response in yeast and mammals. Curr Opin Cell Biol 13:349–355
Pickart CM (2004) Back to the future with ubiquitin. Cell 116:181–190
Pilon M, Schekman R, Romisch K (1997) Sec61p mediates export of a misfolded secretory protein from the endoplasmic reticulum to the cytosol for degradation. EMBO J 16:4540–4548
Plemper RK, Bohmler S, Bordallo J, Sommer T, Wolf DH (1997) Mutant analysis links the translocon and BiP to retrograde protein transport for ER degradation. Nature 388:891–895
Plemper RK, Egner R, Kuchler K, Wolf DH (1998) Endoplasmic reticulum degradation of a mutated ATP-binding cassette transporter Pdr5 proceeds in a concerted action of Sec61 and the proteasome. J Biol Chem 273:32848–32856
Qu D, Teckman JH, Omura S, Perlmutter DH (1996) Degradation of a mutant secretory protein, α1-antitrypsin A, in the endoplasmic reticulum requires proteasome activity. J Biol Chem 271:22791–22795

Rapoport TA, Matlack KE, Plath K, Misselwitz B, Staeck O (1999) Posttranslational protein translocation across the membrane of the endoplasmic reticulum. Biol Chem 380:1143–1150

Ravikumar B, Duden R, Rubinsztein DC (2002) Aggregate-prone proteins with polyglutamine and polyalanine expansions are degraded by autophagy. Hum Mol Genet 11:1107–1117

Ritter C, Helenius A (2000) Recognition of local glycoprotein misfolding by the ER folding sensor UDP-glucose:glycoprotein glucosyltransferase. Nat Struct Biol 7:278–280

Rodighiero C, Tsai B, Rapoport TA, Lencer WI (2002) Role of ubiquitination in retro-translocation of cholera toxin and escape of cytosolic degradation. EMBO Rep 3:1222–1227

Rudiger S, Germeroth L, Schneider-Mergener J, Bukau B (1997) Substrate specificity of the DnaK chaperone determined by screening cellulose-bound peptide libraries. EMBO J 16:1501–1507

Rudiger S, Schneider-Mergener J, Bukau B (2001) Its substrate specificity characterizes the DnaJ co-chaperone as a scanning factor for the DnaK chaperone. EMBO J 20:1042–1050

Russell R, Wali Karzai A, Mehl AF, McMacken R (1999) DnaJ dramatically stimulates ATP hydrolysis by DnaK: insight into targeting of Hsp70 proteins to polypeptide substrates. Biochemistry 38:4165–4176

Rutkowski DT, Kaufman RJ (2004) A trip to the ER: coping with stress. Trends Cell Biol 14:20–28

Sato S, Ward CL, Krouse ME, Wine JJ, Kopito RR (1996) Glycerol reverses the misfolding phenotype of the most common cystic fibrosis mutation. J Biol Chem 271:635–638

Sawkar AR, Cheng WC, Beutler E, Wong CH, Balch WE, Kelly JW (2002) Chemical chaperones increase the cellular activity of N370S beta -glucosidase: a therapeutic strategy for Gaucher disease. Proc Natl Acad Sci U S A 99:15428–15433

Schekman R (2004) Cell biology: a channel for protein waste. Nature 429:817–818

Schmid D, Baici A, Gehring H, Christen P (1994) Kinetics of molecular chaperone action. Science 263:971–973

Schubert U, Anton LC, Gibbs J, Norbury CC, Yewdell JW, Bennink JR (2000) Rapid degradation of a large fraction of newly synthesized proteins by proteasomes. Nature 404:770–774

Shamu CE, Flierman D, Ploegh HL, Rapoport TA, Chau V (2001) Polyubiquitinylation is required for US11-dependent movement of MHC class I heavy chain from endoplasmic reticulum into cytosol. Mol Biol Cell 12:2546–2555

Shen Y, Meunier L, Hendershot LM (2002) Identification and characterization of a novel endoplasmic reticulum (ER) DnaJ homologue, which stimulates ATPase activity of BiP in vitro and is induced by ER stress. J Biol Chem 277:15947–15956

Sifers RN (2003) Cell biology. Protein degradation unlocked. Science 299:1330–1331

Skowronek MH, Hendershot LM, Haas IG (1998) The variable domain of nonassembled Ig light chains determines both their half-life and binding to the chaperone BiP. Proc Natl Acad Sci U S A 95:1574–1578

Song JL, Chuang DT (2001) Natural osmolyte trimethylamine N-oxide corrects assembly defects of mutant branched-chain alpha-ketoacid decarboxylase in maple syrup urine disease. J Biol Chem 276:40241–40246

Spear ED, Ng DT (2003) Stress tolerance of misfolded carboxypeptidase Y requires maintenance of protein trafficking and degradative pathways. Mol Biol Cell 14:2756–2767
Strickland E, Hakala K, Thomas PJ, DeMartino GN (2000) Recognition of misfolding proteins by PA700, the regulatory subcomplex of the 26 S proteasome. J Biol Chem 275:5565–5572
Suh WC, Burkholder WF, Lu CZ, Zhao X, Gottesman ME, Gross CA (1998) Interaction of the Hsp70 molecular chaperone, DnaK, with its cochaperone DnaJ. Proc Natl Acad Sci U S A 95:15223–15228
Tamarappoo BK, Verkman AS (1998) Defective aquaporin-2 trafficking in nephrogenic diabetes insipidus and correction by chemical chaperones. J Clin Invest 101:2257–2267
Taxis C, Hitt R, Park SH, Deak PM, Kostova Z, Wolf DH (2003) Use of modular substrates demonstrates mechanistic diversity and reveals differences in chaperone requirement of ERAD. J Biol Chem 278:35903–35913
Teckman JH, Perlmutter DH (2000) Retention of mutant alpha-antitrypsin Z in endoplasmic reticulum is associated with an autophagic response. Am J Physiol Gastrointest Liver Physiol 279:G961–G974
Teckman JH, Gilmore R, Perlmutter DH (2000) Role of ubiquitin in proteasomal degradation of mutant alpha-antitrypsin Z in the endoplasmic reticulum. Am J Physiol Gastrointest Liver Physiol 278:G39–G48
Teckman JH, Burrows J, Hidvegi T, Schmidt B, Hale PD, Perlmutter DH (2001) The proteasome participates in degradation of mutant alpha 1-antitrypsin Z in the endoplasmic reticulum of hepatoma-derived hepatocytes. J Biol Chem 276:44865–44872
Teckman JH, An JK, Loethen S, Perlmutter DH (2002) Fasting in alpha1-antitrypsin deficient liver: constitutive activation of autophagy. Am J Physiol Gastrointest Liver Physiol 283:G1156–G1165
Thoms S (2002) Cdc48 can distinguish between native and non-native proteins in the absence of cofactors. FEBS Lett 520:107–110
Travers KJ, Patil CK, Wodicka L, Lockhart DJ, Weissman JS, Walter P (2000) Functional and genomic analyses reveal an essential coordination between the unfolded protein response and ER-associated degradation. Cell 101:249–258
Trombetta ES, Parodi AJ (2003) Quality control and protein folding in the secretory pathway. Annu Rev Cell Dev Biol 19:649–676
Tsai B, Ye Y, Rapoport TA (2002) Retro-translocation of proteins from the endoplasmic reticulum into the cytosol. Nat Rev Mol Cell Biol 3:246–255
Vanhove M, Usherwood YK, Hendershot LM (2001) Unassembled Ig heavy chains do not cycle from BiP in vivo but require light chains to trigger their release. Immunity 15:105–114
Varga K, Jurkuvenaite A, Wakefield J, Hong JS, Guimbellot JS, Venglarik CJ, Niraj A, Mazur M, Sorscher EJ, Collawn JF, Bebok Z (2004) Efficient intracellular processing of the endogenous cystic fibrosis transmembrane conductance regulator in epithelial cell lines. J Biol Chem 279:22578–22584
Vashist S, Ng DT (2004) Misfolded proteins are sorted by a sequential checkpoint mechanism of ER quality control. J Cell Biol 165:41–52

Verma R, Chen S, Feldman R, Schieltz D, Yates J, Dohmen J, Deshaies RJ (2000) Proteasomal proteomics: identification of nucleotide-sensitive proteasome-interacting proteins by mass spectrometric analysis of affinity-purified proteasomes. Mol Biol Cell 11:3425–3439

Voges D, Zwickl P, Baumeister W (1999) The 26S proteasome: a molecular machine designed for controlled proteolysis. Annu Rev Biochem 68:1015–1068

Walter J, Urban J, Volkwein C, Sommer T (2001) Sec61p-independent degradation of the tail-anchored ER membrane protein Ubc6p. EMBO J 20:3124–3131

Wang Q, Chang A (1999) Epsl, a novel PDI-related protein involved in ER quality control in yeast. EMBO J 18:5972–5982

Webb JL, Ravikumar B, Atkins J, Skepper JN, Rubinsztein DC (2003) Alpha-Synuclein is degraded by both autophagy and the proteasome. J Biol Chem 278:25009–25013

Weibezahn J, Bukau B, Mogk A (2004) Unscrambling an egg: protein disaggregation by AAA+ proteins. Microb Cell Fact 3:1–12

Weihofen A, Martoglio B (2003) Intramembrane-cleaving proteases: controlled liberation of proteins and bioactive peptides. Trends Cell Biol 13:71–78

Werner ED, Brodsky JL, McCracken AA (1996) Proteasome-dependent endoplasmic reticulum-associated protein degradation: an unconventional route to a familiar fate. Proc Natl Acad Sci U S A 93:13797–13801

Wiertz EJ, Tortorella D, Bogyo M, Yu J, Mothes W, Jones TR, Rapoport TA, Ploegh HL (1996) Sec61-mediated transfer of a membrane protein from the endoplasmic reticulum to the proteasome for destruction. Nature 384:432–438

Wilkinson BM, Tyson JR, Stirling CJ (2001) Sshlp determines the translocation and dislocation capacities of the yeast endoplasmic reticulum. Dev Cell 1:401–409

Wu Y, Swulius MT, Moremen KW, Sifers RN (2003) Elucidation of the molecular logic by which misfolded alpha 1-antitrypsin is preferentially selected for degradation. Proc Natl Acad Sci U S A 100:8229–8234

Ye Y, Meyer HH, Rapoport TA (2003) Function of the p97-Ufd1-Npl4 complex in retrotranslocation from the ER to the cytosol: dual recognition of nonubiquitinated polypeptide segments and polyubiquitin chains. J Cell Biol 162:71–84

Ye Y, Shibata Y, Yun C, Ron D, Rapoport TA (2004) A membrane protein complex mediates retro-translocation from the ER lumen into the cytosol. Nature 429:841–847

Young BP, Craven RA, Reid PJ, Willer M, Stirling CJ (2001) Sec63p and Kar2p are required for the translocation of SRP-dependent precursors into the yeast endoplasmic reticulum in vivo. EMBO J 20:262–271

Yu H, Kopito RR (1999) The role of multiubiquitination in dislocation and degradation of the alpha subunit of the T cell antigen receptor. J Biol Chem 274:36852–36858

Zeitlin PL (2003) Emerging drug treatments for cystic fibrosis. Expert Opin Emerg Drugs 8:523–535

Zhang Y, Nijbroek G, Sullivan ML, McCracken AA, Watkins SC, Michaelis S, Brodsky JL (2001) Hsp70 molecular chaperone facilitates endoplasmic reticulum-associated protein degradation of cystic fibrosis transmembrane conductance regulator in yeast. Mol Biol Cell 12:1303–1314

Zhou M, Schekman R (1999) The engagement of Sec61p in the ER dislocation process. Mol Cell 4:925–934

Zhu X, Zhao X, Burkholder WF, Gragerov A, Ogata CM, Gottesman ME, Hendrickson WA (1996) Structural analysis of substrate binding by the molecular chaperone DnaK. Science 272:1606–1014

CTMI (2006) 300:41–56

CPY* and the Power of Yeast Genetics in the Elucidation of Quality Control and Associated Protein Degradation of the Endoplasmic Reticulum

D. H. Wolf (✉) · A. Schäfer

Institut für Biochemie, Universität Stuttgart, Pfaffenwaldring 55,
70569 Stuttgart, Germany
dieter.wolf@ibc.uni-stuttgart.de

Abstract CPY* is a mutated and malfolded secretory enzyme (carboxypeptidase yscY, Gly255Arg), which is imported into the endoplasmic reticulum but never reaches the vacuole, the destination of its wild type counterpart. Its creation, through mutation, had a major impact on the elucidation of the mechanisms of quality control and associated protein degradation of the endoplasmic reticulum, the eukaryotic organelle, where secretory proteins start the passage to their site of action. The use of CPY* and yeast genetics led to the discovery of a new cellular principle, the retrograde transport of lumenal malfolded proteins across the ER membrane back to their site of synthesis, the cytoplasm. These tools furthermore paved the way for our current understanding of the basic mechanism of malfolded protein discovery in the ER and their ubiquitin-proteasome driven elimination in the cytosol (ERQD).

1
CPY*, a Malfolded Secretory Protein Is Retained in the Endoplasmic Reticulum and Degraded in the Cytosol

During the establishment of the yeast *Saccharomyces cerevisiae* as a model organism to study the function of proteolysis in eukaryotic cell physiology via biochemical and genetic means, the first protease mutant defective in the activity of one of the vacuolar proteases, carboxypeptidase yscY (CPY), was isolated [79]. As a protein of the hydrolytic vacuolar (lysosomal) compartment, CPY is synthesized in the cytosol as a pre-pro-enzyme and thereafter enters the secretory pathway: after import into the endoplasmic reticulum (ER) the pre-(signal)-sequence is cleaved off and the enzyme is folded. During these processes, disulfide bonds are formed and CPY is modified with four N-linked carbohydrate chains, yielding p1-CPY. After outer chain mannosylation in the Golgi-apparatus, CPY enters the vacuole where the pro-sequence is cleaved to yield the mature form of the enzyme [64, 47]. The fact that mutated CPY had never matured to the vacuolar wild type form indicated that the mutation had either destroyed the maturation site of this serine protease, thus preventing its cleavage into the mature form in the vacuole, or that the mutation prohibited secretion of the protein to its vacuolar location [51]. Sequencing of the mutant gene uncovered that a mutation, Gly255Arg, in a highly conserved site of all serine proteases, two amino acids away from the active site serine, had occurred [20]. Incubation of the mutant protein with trypsin in vitro leads to its rapid degradation, in contrast to wild type pro-CPY, which is cleaved to its mature size. This indicates that the mutant protein is completely differently folded as compared to the wild type protein. This malfolded pro-CPY protein was named CPY* [20]. Even though CPY* contains the pro-sequence, which could direct it to the vacuole, it never reaches this organelle. This was very surprising at the time, as on one hand the vacuole is the working place of CPY, on the other hand it represents the gut of the cell responsible for degrading cellular proteins in an unspecific way. Instead, CPY* is retained in the endoplasmic reticulum (ER) and rapidly degraded, with a half life of 15–20 min [20]. Two developments, which had merged in 1991, made the subsequent discoveries possible: (a) the elucidation of the ubiquitin system, which through tagging selected proteins with the 76-amino acid protein ubiquitin, targets them for degradation [72, 29] and (b) the discovery of the proteasome as the proteolytic machinery, which degrades the ubiquitin-tagged proteins in vivo [27, 78, 80]. The establishment of yeast in the elucidation of the physiological function of vacuolar [77, 70, 78] and ubiquitin-proteasome linked proteolysis [32, 72] and the finding of CPY* as a rapidly degraded, malfolded secretory protein [20, 65, 61] were crucially important for the dissection of ER quality control and the associated cytosolic degradation pathway (ERQD).

2 Endoplasmic Reticulum to Cytosol Retrotranslocation: A New Cellular Mechanism

The isolation of yeast mutants defective in the degradation of CPY* started to give crucial insights into the quality control and degradation pathway [45]. After mutagenesis of a yeast strain carrying the *prc1-1* allele encoding CPY*, mutants were isolated on the basis of defective CPY* degradation, which was made visible on colony immunoblots of mutated strains using CPY* antibodies [20, 45]. The first series of seven mutant alleles giving rise to a disturbed ER quality control and degradation process of CPY* were named *der1* to *der7* (der, degradation of the ER) [45, 37]. Analysis of the *der2* mutant led to a breakthrough in our knowledge of how ER-associated degradation works. The *DER2* gene was identified as the gene encoding the ubiquitin-conjugating enzyme (E2) Ubc7p [31]. The participation of an ubiquitin-conjugating enzyme in the degradation of CPY* immediately pointed to the participation of the ubiquitin-proteasome system (UPS) in the degradation of the malfolded enzyme. Analysis of CPY* degradation in proteasome mutants indeed confirmed the requirement for the 26S proteasome in the degradation process [31]. As the 26S proteasome had only been found in the cytosol and the nucleus of cells [44], and Ubc7p only in the cytosol [42], a retrograde transport of CPY* from the ER lumen back to the cytoplasm had to be postulated. The appearance of polyubiquitinated and at the same time glycosylated CPY* on the cytoplasmic side of the ER membrane substantiated this idea [31]: CPY* had obviously been imported into the ER and N-glycosylated, somehow recognized as being unable to fold, transported back out of the ER and polyubiquitinated for subsequent proteasomal degradation (Fig. 1). These findings had set the stage for a new biological mechanism. It violated the dogma that proteins, which had entered the endoplasmic reticulum, were trapped in the secretory pathway unable to return back into the cytoplasm [5]. The findings of Hiller et al. in 1996 [31], using CPY* as a substrate, had set the frame around a mosaic composed of two processes: (a) protein quality control in the endoplasmic reticulum and (b) degradation of malfolded proteins in the cytoplasm via the ubiquitin-proteasome system. Both processes are linked via a transport step of the malfolded protein back out of the ER into the cytoplasm. Additional work was and still is required to fill the gaps in knowledge in the mosaic. Here also CPY* continues to serve as an excellent model substrate. By introducing a fifth glycosylation site at the very C-terminus in addition to the four N-glycosylation sites of CPY*, it could be shown that this CPY* molecule receives five carbohydrates. This CPY*derivative was found to be degraded as the authentic CPY* molecule. As glycosylation occurs 12–14 amino acids away from the translocon in the lumen of the ER [54], the CPY* molecule containing the fifth glycosylation at the very C-terminus must have entered the ER lumen completely prior to its degradation via the ubiquitin-proteasome system [59]. Thus, a targeting of ER lumenal CPY* back to some translocation channel must occur, followed by retrotranslocation and, after polyubiquitination, degradation by the proteasome (Fig. 1).

3 Carbohydrate Trimming: A Tool of the Endoplasmic Reticulum Quality Control of Glycoproteins

The lumen of the ER contains a highly active protein quality control machinery [18, 17]. For N-glycosylated proteins, trimming of the $Glc_3Man_9GlcNAc_2$ oligosaccharides is an important process of the quality control mechanism. The N-glycans are matured by stepwise removal of the three terminal glucose residues via α-glucosidases I and II. Finally, α1,2-mannosidase I releases a mannose residue from the inner branch of the N-glycan, giving rise to $Man_8GlcNAc_2$ [28, 17]. It is thought that this process sets the timer for folding, and when unsuccessful, for degradation of the N-glycosylated secretory protein. When the mannose-9 residue is cleaved off the $Man_9GlcNAc_2$ structure, the protein is retained in the ER and delivered for elimination. Mutant analysis using CPY* as malfolded protein has shown that indeed α-glucosidase I, found in the *der* screen as Der7p [37], and glucosidase II [40] as well as α1,2-mannosidase I [46] are required for degradation of the CPY* protein, thus substantiating the proposed quality control mechanism of N-glycosylated proteins (Fig. 1). In a systematic study of the four N-glycans of CPY* in ERQD, substantial differences in their signaling function were found: of the four N-linked carbohydrate chains at positions Asn13, Asn87, Asn168, and Asn368, only the presence of the Asn368-linked glycan is necessary and sufficient for efficient degradation of CPY* [49a, 65a]. Recent studies have shown that degradation of CPY* also requires the lectin-like protein Htm1p/Mnl1p [39, 53]. It is proposed that Htm1p/Mnl1p recognizes the trimmed $Man_8GlcNAc_2$ structure of CPY* and other N-glycosylated malfolded proteins, retaining them in the ER and finally delivering them to proteasomal degradation.

One of the *DER* genes required for CPY* degradation is *DER5*, encoding the Ca^{2+}/Mn^{2+} pump Pmr1p: a *DER5* deletion considerably slows down CPY* degradation [16]. Pmr1p is localized to the Golgi but is also required for maintaining normal Ca^{2+} levels in the ER [66]. A second Ca^{2+} pump required for undisturbed degradation of CPY* was found to be Cod1p, localized in the ER membrane [73]. Both Pmr1p and Cod1p are required for Ca^{2+} homeostasis of

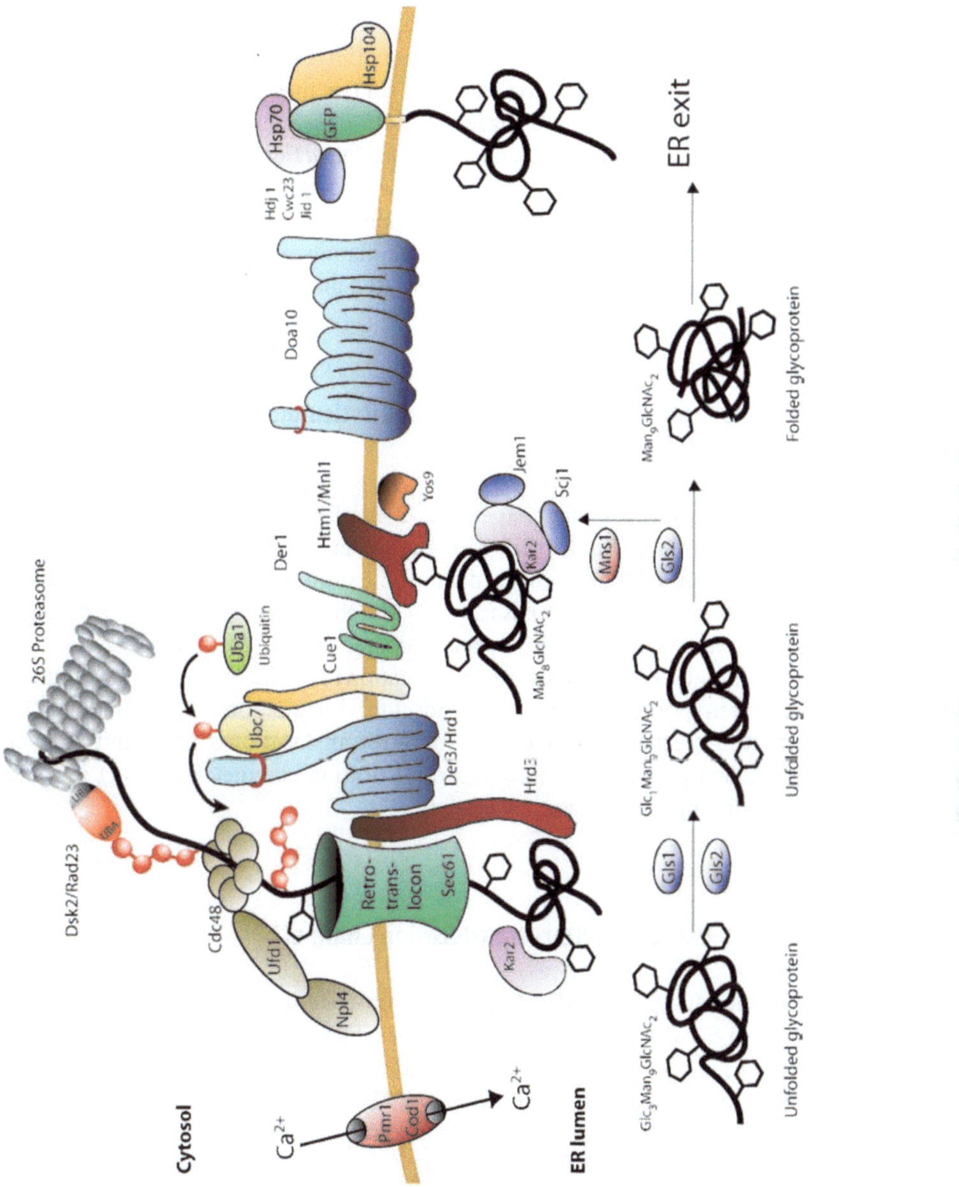

Fig. 1 Protein quality control and elimination of misfolded endoplasmic reticulum proteins. (Modified after Kostova and Wolf, 2003) [49]

the ER [13, 73]. In contrast to CPY*, degradation of nonglycosylated ER substrates was not affected by the absence of both Ca^{2+} pumps; Ca^{2+}homeostasis in the ER must be linked to the quality control mechanism of N-glycosylated proteins [73]. α1,2-Mannosidase I is a Ca^{2+}-dependent enzyme. Indeed, analysis of N-linked oligosaccharides in Δ*pmr1*Δ*cod1* double mutants uncovered a large portion of protein-linked sugar being of the untrimmed $Man_9GlcNAc_2$ type [73]. This indicates that one function of ER-calcium in the degradation of CPY* rests in its ability to render α1,2-mannosidase I active and thus allow proper ER quality control of N-glycosylated proteins.

4 Soluble Proteins Require Endoplasmic Reticulum-Lumenal Chaperones for Degradation

The search for lumenal chaperones of the ER required for degradation of CPY* uncovered Kar2p (BiP in mammalian cells) [57]. Besides CPY*, the degradation of other soluble mutated ER proteins such as mutated pro-α-factor is also dependent on Kar2p [10] and the interacting DnaJ-like proteins Jem1p and Scj1p [55]. It is proposed that the Kar2p machinery prevents soluble malfolded proteins from aggregation in the ER lumen, thus facilitating their retrograde export into the cytosol for degradation [55, 69] (Fig. 1). The targeting mechanism responsible for retrotranslocation of CPY* and all other malfolded proteins to some retrotranslocation channel for export into the cytosol is not yet known.

5 Sec61p, Part of the Retrotranslocation Channel?

Mutant studies using CPY* as ERQD substrate indicate that the translocon protein Sec61p, which forms the import channel for secretory proteins into the ER, is also part of the export channel delivering CPY* to the cytosolic ubiquitin-proteasome machinery [57]. Also genetic interaction studies point to Sec61p as being part of the CPY* export channel [58]. These studies also indicate a composition of the export channel, which is different from the import channel (Fig. 1). Studies on a variety of other ERQD proteins point to the participation of Sec61p in the retrotranslocation process as well [76, 56, 3; 14]. The appearance of glycosylated CPY* in the cytoplasm [31, 41] points to a diameter of the retrotranslocation channel, which must be larger than the pore size of the import channel, which has only to accommodate a single polypeptide chain. However, recent experimentation seems to indicate that the use of Sec61p in retrotranslocation may not be a unique principle [38, 82]. Final proof for the nature of a retrotranslocation channel will only come from isolation of such a channel in the process of protein export.

6
Endoplasmic Reticulum-Associated Protein Degradation: Ubiquitin, the Proteasome and Other Helpers

Retrotranslocation of CPY*—and the majority of proteins destined for endoplasmic reticulum-associated protein degradation (ERAD) [9, 61, 48, 49, 34]—is followed by polyubiquitination and proteasomal degradation. The biochemical search for the number of ubiquitin-conjugating enzymes (Ubcs) that had overlapping specificity with the *DER2* gene product Ubc7p in the polyubiquitination process of CPY* uncovered two additional members: Ubc6p [31], an integral ER membrane localized E2, the active site facing the cytoplasm [74], and Ubc1p, a cytosolic E2 [21]. Of the three ubiquitin-conjugating enzymes, Ubc7p has the strongest influence on the degradation of CPY*. Interestingly, the soluble cytoplasmically localized Ubc7p gains its activity for polyubiquitinating CPY* and other proteins only after binding to an ER membrane protein, Cue1p [4].

On the basis of the mutant screen using CPY* for the discovery of components of the ubiquitination and degradation machinery, Der3p/Hrd1p was uncovered [6]. It is a polytopic ER membrane protein containing six transmembrane domains with its N- and C-terminus facing the cytoplasm [15]. It contains a RING-H2 finger domain in its C-terminus [7] and turned out to be the ubiquitin-protein ligase (E3) polyubiquitinating CPY* [15] as well as other proteins destined for ER degradation via the proteasome [1]. In a search for mutants defective in regulated degradation of the ER membrane-located enzyme hydroxymethylglutaryl (HMG) CoA reductase, an additional membrane protein, named Hrd3p, was found, which interacts with Der3p/Hrd1p [25, 23, 15] and which was also found to be required for the degradation of CPY* [58]. It is thought to be a device signaling the presence of malfolded proteins in the ER to the cytoplasmically located ubiquitination and degradation machinery [23, 15]. Der3p/Hrd1p represents an ubiquitin-protein ligase, which is responsible for polyubiquitination of a certain set of malfolded ER proteins, among them CPY* (Fig. 1). A second, polytopic ER membrane-located ubiquitin-protein ligase is Doa10p [68], which is responsible for polyubiquitination of a different set of ER proteins destined for proteasomal degradation [75, 24, 38].

As the 26S proteasome contains six different ATPases in the base of its 19S cap (regulator) complex, it was thought that pulling of the ubiquitinated malfolded proteins away from the ER and delivering them to the 20S proteasome core complex for degradation was carried out by these ATPase subunits. Using CPY* as substrate, four research groups at nearly the same time uncovered that transport of the malfolded substrate from the ER to the proteasome requires the AAA-ATPase Cdc48 (p97 in mammals) and two additional complexing proteins, Ufd1p and Npl4p [2, 81, 41, 62]. Mutations in the components of this trimeric complex lead to a failure of delivery of CPY* into the cytosol, leaving ubiquitinated CPY* bound to the ER [41]. The requirement of the trimeric Cdc48 complex for ERAD has been shown for all tested proteins so far, which require polyubiquitination for degradation [2, 81, 8; 62, 24, 38] (Fig. 1).

7 Modular CPY*-Based Membrane Substrates Broaden the Picture

The use of CPY* and different other substrates during time to study ERQD had given insight into a basic machinery, which was equally necessary for elimination of all substrates tested. For glycosylated proteins, this machinery constitutes of the glucosidases I and II, α-mannosidase I, and Htm1p/Mnl1p (EDEM) for quality control assessment. On the cytoplasmic side of the ER membrane, the ubiquitin-conjugating enzyme Ubc7p, depending on the substrate, either Der3/Hrd1p or Doa10p as ubiquitin-protein ligases, the trimeric AAA-ATPase complex Cdc48 (p97)-Ufd1-Npl4p, and the 26S proteasome, were shown to be required for the degradation of all substrates tested so far (Fig. 1). There was a discrepancy in the use of ER-lumenal and cytoplasmic chaperones for degradation of malfolded soluble and membrane proteins [57, 60, 10, 30, 55, 83] and in the use of an ER membrane protein, Der1p, required for CPY* ERAD but not membrane protein ERAD [45, 60, 30, 36]. The construction of three topologically different modular substrates all containing CPY* as the malfolded protein in the lumen of the ER shed more light on the question of which ERQD system components are generally used for recognition and degradation of topologically different proteins containing the same malfolded recognition domain. The set of molecules used consisted of CPY*, a CPY* molecule linked to a transmembrane domain (CT*), and transmembrane-linked CPY* containing a strongly folding cytoplasmic domain, the green-fluorescent protein (GFP) (CTG*) [69] (Fig. 2). As previously found for several completely different substrate species containing different malfolded domains, the basic machinery required for degradation of the three topologically different CPY* substrates (CPY*, CT*, CTG*)

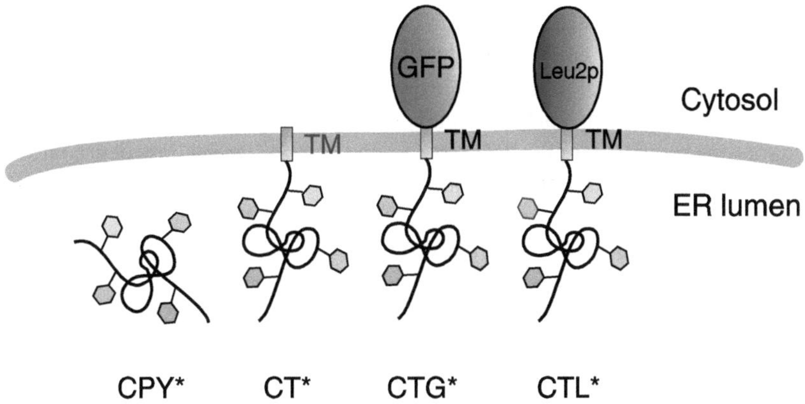

Fig. 2 Schematic presentation of CPY* and its variants

consisted of the ubiquitin conjugating enzyme Ubc7p, the ubiquitin-protein ligase Der3/Hrd1p, the trimeric Cdc48-Ufd1-Npl4p complex, and the 26S proteasome [69]. The ER lumenal Hsp70 chaperone Kar2p was only required for degradation of soluble CPY* and not for degradation of any of the membrane-bound CPY* species [69]. When testing the requirement of cytoplasmic chaperones for degradation of the topologically different CPY* species, it turned out that degradation of CTG* containing the tightly folding GFP domain was crucially dependent on the presence of the cytoplasmic Hsp70 chaperones of the Ssa family. Neither degradation of soluble CPY* nor of membrane bound CPY* without cytoplasmic domain required Ssa1p. The most likely explanation for this finding is that unfolding of the tightly folding GFP domain of CTG* is dependent on these cytoplasmic Hsp70 chaperones to allow ER removal and/or proteasomal degradation of this protein species. A less but clearly observable effect on degradation of CTG* was also found for the DnaJ orthologs Hlj1p, Cwc23p, and Jid1p as well as for the Hsp100 member of chaperones, Hsp104 [69]. The integral ER membrane protein Der1p was only required for degradation of soluble CPY* and not for degradation of any other membrane-bound CPY* species [69]. The finding of an interaction of a mammalian ortholog of Der1p, Derlin-1, with membrane-bound MHC class I molecules in association with the viral US11 protein for degradation upon cytomegalovirus infection of cells is therefore somewhat surprising [82]. One may assume that recruitment of the Der1 ortholog to a membrane protein is virus specific, and may even be specific for the US11 protein: the second cytomegalovirus protein, US2, which also targets MHC class I heavy chains for degradation [76], does not seem to work in conjunction with Derlin-1 [50].

For the moment, one may conclude that a machinery exists for ERAD of all misfolded proteins, which indeed consists of the ubiquitin-conjugating enzyme Ubc7p, the ubiquitin protein ligase Der3/Hrd1p and/or Doa10p, the trimeric Cdc48-Ufd1-Npl4p complex, and the 26S proteasome (Fig. 1). The chaperone requirement for degradation may vary, whereby only degradation of soluble proteins seems to require ER lumenal Kar2p/BiP. Also, the use of additional ubiquitin conjugating enzymes as Ubc1p or Ubc6p may vary from substrate to substrate. Obviously the use of the ubiquitin-protein ligases Der3/Hrd1p and Doa10p is also substrate-dependent but not necessarily exclusive for one or the other malfolded ER protein [24, 38]. Using CPY* as ERQD substrate, the requirement of a cytoplasmic peptide: N-glycanase (PNGase), for undisturbed proteasomal degradation of the malfolded protein was shown [67]. It is thought that PNGase cleaves off the carbohydrate residues from N-glycosylated ERQD substrates to allow their efficient elimination by the proteasome [33, 35].

8
Yeast Genomics Discovers New Players

The elucidation of the yeast genome with its roughly 6,000 open reading frames had been followed by the construction of yeast deletion libraries, which consist of approximately 5,000 individual clones each deleted for a single nonessential gene (i.e., EUROSCARF deletion library, Frankfurt, Germany). Such a deletion library can be used for genome-wide screens of mutants defective in a certain cellular process. As cells can tolerate a defect in ERQD as long as the unfolded protein response (UPR) is intact (CPY* and other misfolded proteins induce the UPR) [45, 21, 71], the existence of such a deletion library made it possible to search for new components of ERQD by performing a genomic screen. As cell growth is one of the most sensitive indicators of alterations in cell physiology due to mutations, a growth test on the basis of degradation of CPY* was developed. The fact that CTG*, carrying CPY* in the ER lumen and GFP in the cytosol behaved as an excellent ERQD substrate, the cytosolic GFP domain was exchanged for the Leu2 protein (3-isopropylmalate-dehydrogenase), leading to the new substrate CTL* [11, 52] (Fig. 2). Cells carrying a *LEU2* deletion can only grow when CTL* is present, the cytoplasmic Leu2p domain of the fusion protein being able to complement the leucine auxotrophy of cells: Strains with *leu2* auxotrophy but otherwise wild type for ERQD are unable to grow in media lacking leucine. Only when ERQD is defective is CTL* stabilized and able to complement the *LEU2* deficiency and thus allow growth [11, 52]. Screening of the nearly 5,000 individual deletion mutants of the EUROSCARF deletion library expressing CTL* resulted in finding most of the known ERQD components [52]. In addition, however, the search resulted in the discovery of a variety of new mutants defective in ERQD. Among these were mutants deleted in genes of the ubiquitin-like (UBL) and ubiquitin-associated (UBA) domain proteins Rad23p and Dsk2p as well as the mannose-6-phosphate receptor-like domain protein Yos9p [11, 52]. Interestingly Yos9p is only required for degradation of N-glycosylated proteins, not for degradation of nonglycosylated ERQD substrates. Its localization in the ER may lead to the proposal that Yos9p is a lectin-like protein, which acts in concert with or sequentially with Htm1p/Mnl1p in the quality control process of N-glycosylated proteins [11]. Dsk2p and Rad23p were shown to be able to bind ubiquitinated proteins via their UBA domain and dock onto the 19S cap subunit of the proteasome via their UBL domain [12, 22, 63, 19, 26, 43]. In contrast to mutants defective in the trimeric Cdc48 complex in which polyubiquitinated CPY* remains bound to the ER, in Δ*dsk2*Δ*rad23* double mutants, a substantial amount of polyubiquitinated CPY* is found in the soluble fraction of cells. At the moment, the most plausible explanation for Dsk2p and Rad23p action in ERAD is their function as receptors shuttling polyubiquitinated ERQD substrates from the trimeric Cdc48 complex to the proteasome. By keeping malfolded ER lumenal and especially malfolded ER membrane proteins with their exposed hydrophobic domains complexed to soluble partners in the cytoplasm until their degradation in the proteasome, the cell avoids aggregation and precipitation of these malfolded proteins in the cytoplasm, preventing diseased cell states. CPY* and yeast genetics have indeed paved the way to our understanding of the basic mechanisms of protein quality control of the ER and ER-associated degradation (Fig. 1). They will certainly continue.

References

1. Bays NW et al (2001) Hrd1p/Der3p is a membrane-anchored ubiquitin ligase required for ER- associated degradation. Nat Cell Biol 3:24–29
2. Bays NW et al (2001) HRD4/NPL4 Is required for the proteasomal processing of ubiquitinated ER proteins. Mol Biol Cell 12:4114–4128
3. Bebök Z et al (1998) The mechanism underlying cystic fibrosis transmembrane conductance regulator transport from the endoplasmic reticulum to the proteasome includes Sec61beta and a cytosolic, deglycosylated intermediary. J Biol Chem 273:29873–29878
4. Biederer T et al (1997) Role of Cue1p in ubiquitination and degradation at the ER surface. Science 278(5344):1806–1809
5. Blobel G (1995) Unidirectional and bidirectional protein traffic across membranes. Cold Spring Harb Symp Quant Biol 60:1–10

6. Bordallo J et al (1998) Der3p/Hrd1p is required for endoplasmic reticulum-associated degradation of misfolded lumenal and integral membrane proteins. Mol Biol Cell 9:209–222
7. Bordallo J, Wolf DH (1999) A RING-H2 finger motif is essential for the function of Der3/Hrd1 in endoplasmic reticulum associated protein degradation in the yeast Saccharomyces cerevisiae. FEBS Lett 448:244–248
8. Braun S et al (2002) Role of the ubiquitin-selective CDC48(UFD1/NPL4) chaperone (segregase) in ERAD of OLE1 and other substrates. EMBO J 21:615–621
9. Brodsky JL, McCracken AA(1999) ER protein quality control and proteasome-mediated protein degradation. Semin Cell Dev Biol 10:507–513
10. Brodsky JL et al (1999) The requirement for molecular chaperones during endoplasmic reticulum-associated protein degradation demonstrates that protein export and import are mechanistically distinct. J Biol Chem 274:3453–3460
11. Buschhorn B et al (2004) A genome wide screen identifies Yos9p as a new lectin essential for ER-associated degradation (ERAD) of glycoproteins. FEBS Lett 577:422–426
12. Chen L, Madura K (2002) Rad23 promotes the targeting of proteolytic substrates to the proteasome. Mol Cell Biol 22:4902–4913
13. Cronin SR et al (2002) Cod1p/Spf1p is a P-type ATPase involved in ER function and Ca2+ homeostasis. J Cell Biol 157:1017–1028
14. De Virgilio M et al (1998) Ubiquitination is required for the retro-translocation of a short-lived luminal endoplasmic reticulum glycoprotein to the cytosol for degradation by the proteasome. J Biol Chem 273:9734–9743
15. Deak PM, Wolf DH (2001) Membrane topology and function of Der3/Hrd1p as a ubiquitin-protein ligase (E3) involved in endoplasmic reticulum degradation. J Biol Chem 276:10663–10669
16. Dürr G et al (1998) The medial-Golgi ion pump Pmr1 supplies the yeast secretory pathway with Ca2+ and Mn2+ required for glycosylation, sorting, and endoplasmic reticulum-associated protein degradation. Mol Biol Cell 9:1149–1162
17. Ellgaard L, Helenius A (2003) Quality control in the endoplasmic reticulum. Nat Rev Mol Cell Biol 4:181–191
18. Ellgaard L et al (1999) Setting the standards: quality control in the secretory pathway. Science 286(5446):1882–1888
19. Elsasser S et al (2004) Rad23 and Rpn10 serve as alternative ubiquitin receptors for the proteasome. J Biol Chem 279:26817–26822
20. Finger A et al (1993) Analysis of two mutated vacuolar proteins reveals a degradation pathway in the endoplasmic reticulum or a related compartment of yeast. Eur J Biochem 218:565–574
21. Friedländer R et al (2000) A regulatory link between ER-associated protein degradation and the unfolded-protein response. Nat Cell Biol 2:379–384
22. Funakoshi M et al (2002) Budding yeast Dsk2p is a polyubiquitin-binding protein that can interact with the proteasome. Proc Natl Acad Sci U S A 99:745–750
23. Gardner RG et al (2000) Endoplasmic reticulum degradation requires lumen to cytosol signaling. Transmembrane control of Hrd1p by Hrd3p. J Cell Biol 151:69–82

24. Gnann A et al (2004) Cystic fibrosis transmembrane conductance regulator degradation depends on the lectins Htm1p/EDEM and the Cdc48 protein complex in yeast. Mol Biol Cell 15:4125–4135
25. Hampton RY et al (1996) Role of 26S proteasome and HRD genes in the degradation of 3-hydroxy-3- methylglutaryl-CoA reductase, an integral endoplasmic reticulum membrane protein. Mol Biol Cell 7:2029–2044
26. Hartmann-Petersen R, Gordon C (2004) Proteins interacting with the 26S proteasome. Cell Mol Life Sci 61:1589–1595
27. Heinemeyer W et al (1991) Proteinase yscE, the yeast proteasome/multicatalytic-multifunctional proteinase: mutants unravel its function in stress induced proteolysis and uncover its necessity for cell survival. EMBO J 10:555–562
28. Helenius A, Aebi M (2001) Intracellular functions of N-linked glycans. Science 291(5512):2364–2369
29. Hershko A, Ciechanover A (1998) The ubiquitin system. Annu Rev Biochem 67:425–479
30. Hill K, Cooper AA (2000) Degradation of unassembled Vph1p reveals novel aspects of the yeast ER quality control system. EMBO J 19:550–561
31. Hiller MM et al (1996) ER degradation of a misfolded luminal protein by the cytosolic ubiquitin-proteasome pathway. Science 273(5282):1725–1728
32. Hilt W, Wolf DH (1996) Proteasomes: destruction as a programme. Trends Biochem Sci 21:96–102
33. Hirsch C et al (2003) A role for N-glycanase in the cytosolic turnover of glycoproteins. EMBO J 22:1036–1046
34. Hirsch C et al (2004) Endoplasmic reticulum associated protein degradation—one model fits all? Biochim Biophys Acta 1695:215–223
35. Hirsch C et al (2004) Yeast N-glycanase distinguishes between native and non-native glycoproteins. EMBO Rep 5:201–206
36. Hitt R, Wolf DH (2004) Der1p, a protein required for degradation of malfolded soluble proteins of the endoplasmic reticulum: topology and Der1-like proteins. FEMS Yeast Res 4:721–729
37. Hitt R, Wolf DH (2004) DER7, encoding alpha-glucosidase I is essential for degradation of malfolded glycoproteins of the endoplasmic reticulum. FEMS Yeast Res 4:815–820
38. Huyer G et al (2004) Distinct machinery is required in Saccharomyces cerevisiae for the endoplasmic reticulum-associated degradation of a multispanning membrane protein and a soluble luminal protein. J Biol Chem 279:38369–38378
39. Jakob CA et al (2001) Htm1p, a mannosidase-like protein, is involved in glycoprotein degradation in yeast. EMBO Rep 2:423–430
40. Jakob CA et al (1998) Degradation of misfolded endoplasmic reticulum glycoproteins in Saccharomyces cerevisiae is determined by a specific oligosaccharide structure. J Cell Biol 142:1223–1233
41. Jarosch E et al (2002) Protein dislocation from the ER requires polyubiquitination and the AAA- ATPase Cdc48. Nat Cell Biol 4:134–139
42. Jungmann J et al (1993) Resistance to cadmium mediated by ubiquitin-dependent proteolysis. Nature 361(6410):369–371
43. Kim I et al (2004) Multiple interactions of rad23 suggest a mechanism for ubiquitylated substrate delivery important in proteolysis. Mol Biol Cell 15:3357–3365

44. Knecht E, Rivett AJ (2000). Intracellular localization of proteasomes. In: Hilt W, Wolf DH (eds) Proteasomes: the world of regulatory proteolysis. Landes Bioscience, Georgetown/Eurekah.com, Austin, TX, pp 176–185
45. Knop M et al (1996) Der1, a novel protein specifically required for endoplasmic reticulum degradation in yeast. EMBO J 15:753–763
46. Knop M et al (1996) N-Glycosylation affects endoplasmic reticulum degradation of a mutated derivative of carboxypeptidase yscY in yeast. Yeast 12:1229–1238
47. Knop M et al (1993) Vacuolar/lysosomal proteolysis: proteases, substrates, mechanisms. Curr Opin Cell Biol 5:990–996
48. Kostova Z, Wolf DH (2002). Protein quality control in the export pathway: the endoplasmic reticulum and its cytoplasmic proteasome connection. In: Dalbey RE, von Heijne G (eds) Protein targeting, transport and translocation. London, Academic Press, pp 180–213
49. Kostova Z, Wolf DH (2003) For whom the bell tolls: protein quality control of the endoplasmic reticulum and the ubiquitin-proteasome connection. EMBO J 22:2309–2317
49a. Kostova Z, Wolf DH (2005) Importance of carbohydrate positioning in the recognition of mutated CPY for ER-associated degradation. J Cell Science 118:1485–1492
50. Lilley BN, Ploegh HL (2004) A membrane protein required for dislocation of misfolded proteins from the ER. Nature 429(6994):834–840
51. Mechler B et al (1982) In vivo biosynthesis of vacuolar proteinases in proteinase mutants of Saccharomyces cerevisiae. Biochem Biophys Res Commun 107:770–778
52. Medicherla B et al (2004) A genomic screen identifies Dsk2p and Rad23p as essential components of ER-associated degradation. EMBO Rep 5:692–697
53. Nakatsukasa K et al (2001) Mnl1p, an alpha -mannosidase-like protein in yeast Saccharomyces cerevisiae, is required for endoplasmic reticulum-associated degradation of glycoproteins. J Biol Chem 276:8635–8638
54. Nilsson IM, G von Heijne (1993) Determination of the distance between the oligosaccharyltransferase active site and the endoplasmic reticulum membrane. J Biol Chem 268:5798–5801
55. Nishikawa S et al (2001) Molecular chaperones in the yeast endoplasmic reticulum maintain the solubility of proteins for retrotranslocation and degradation. J Cell Biol 153:1061–1070
56. Pilon M et al (1997) Sec61p mediates export of a misfolded secretory protein from the endoplasmic reticulum to the cytosol for degradation. EMBO J 16:4540–4548
57. Plemper RK et al (1997) Mutant analysis links the translocon and BiP to retrograde protein transport for ER degradation. Nature 388(6645):891–895
58. Plemper RK et al (1999) Genetic interactions of Hrd3p and Der3p/Hrd1p with Sec61p suggest a retro-translocation complex mediating protein transport for ER degradation. J Cell Sci 112:4123–4134
59. Plemper RK et al (1999) Re-entering the translocon from the lumenal side of the endoplasmic reticulum. Studies on mutated carboxypeptidase yscY species. FEBS Lett 443:241–245
60. Plemper RK et al (1998) Endoplasmic reticulum degradation of a mutated ATP-binding cassette transporter Pdr5 proceeds in a concerted action of Sec61 and the proteasome. J Biol Chem 273:32848–32856

61. Plemper RK, Wolf DH (1999) Retrograde protein translocation: ERADication of secretory proteins in health and disease. Trends Biochem Sci 24:266–270
62. Rabinovich E et al (2002) AAA-ATPase p97/Cdc48p, a cytosolic chaperone required for endoplasmic reticulum-associated protein degradation. Mol Cell Biol 22:626–634
63. Rao H, Sastry A (2002) Recognition of specific ubiquitin conjugates is important for the proteolytic functions of the ubiquitin-associated domain proteins Dsk2 and Rad23. J Biol Chem 277:11691–11695
64. Rendueles PS, Wolf DH (1988) Proteinase function in yeast: biochemical and genetic approaches to a central mechanism of post-translational control in the eukaryote cell. Fems Microbiol Rev 4:17–45
65. Sommer T, Wolf DH (1997) Endoplasmic reticulum degradation: reverse protein flow of no return. FASEB J 11:1227–1233
65a. Spear ED, Ng DTW (2005) Single, context-specific glycans can target misfolded glycoproteins for ER-associated degradation. J Cell Biol 169:73–82
66. Strayle J et al (1999) Steady-state free Ca(2+) in the yeast endoplasmic reticulum reaches only 10 microM and is mainly controlled by the secretory pathway pump pmr1. EMBO J 18:4733–4743
67. Suzuki T et al (2000) PNG1, a yeast gene encoding a highly conserved peptide: N-glycanase. J Cell Biol 149:1039–1052
68. Swanson R et al (2001) A conserved ubiquitin ligase of the nuclear envelope/endoplasmic reticulum that functions in both ER-associated and Matalpha2 repressor degradation. Genes Dev 15:2660–2674
69. Taxis C et al (2003) Use of modular substrates demonstrates mechanistic diversity and reveals differences in chaperone requirement of ERAD. J Biol Chem 278:35903–35913
70. Teichert U et al (1989) Lysosomal (vacuolar) proteinases of yeast are essential catalysts for protein degradation, differentiation, and cell survival. J Biol Chem 264:16037–16045
71. Travers KJ et al (2000) Functional and genomic analyses reveal an essential coordination between the unfolded protein response and ER-associated degradation. Cell 101:249–258
72. Varshavsky A (1997) The ubiquitin system. Trends Biochem Sci 22:383–387
73. Vashist S et al (2002) Two distinctly localized p-type ATPases collaborate to maintain organelle homeostasis required for glycoprotein processing and quality control. Mol Biol Cell 13:3955–3966
74. Walter J et al (2001) Sec61p-independent degradation of the tail-anchored ER membrane protein Ubc6p. EMBO J 20:3124–3131
75. Wang Q, Chang A (2003) Substrate recognition in ER-associated degradation mediated by Eps1, a member of the protein disulfide isomerase family. EMBO J 22:3792–3802
76. Wiertz EJ et al (1996) Sec61-mediated transfer of a membrane protein from the endoplasmic reticulum to the proteasome for destruction. Nature 384(6608):432–438
77. Wolf DH (1982) Proteinase action in vitro versus proteinase function in vivo: mutants shed light on intracellular proteolysis in yeast. Trends Biochem Sci 7:35–37

78. Wolf DH (2004) Ubiquitin-proteasome system: from lysosome to proteasome: the power of yeast in the dissection of proteinase function in cellular regulation and waste disposal. Cell Mol Life Sci 61:1601–1614
79. Wolf DH, Fink GR (1975) Proteinase C (carboxypeptidase Y) mutant of yeast. J Bacteriol 123:1150–1156
80. Wolf DH, Hilt W (2004) The proteasome: a proteolytic nanomachine of cell regulation and waste disposal. Biochim Biophys Acta 1695:19–31
81. Ye Y et al (2001) The AAA ATPase Cdc48/p97 and its partners transport proteins from the ER into the cytosol. Nature 414(6864):652–656
82. Ye Y et al (2004) A membrane protein complex mediates retro-translocation from the ER lumen into the cytosol. Nature 429(6994):841–847
83. Zhang Y et al (2001) Hsp70 molecular chaperone facilitates endoplasmic reticulum-associated protein degradation of cystic fibrosis transmembrane conductance regulator in yeast. Mol Biol Cell 12:1303–1314

CTMI (2006) 300:57–93

The Role of the Ubiquitination Machinery in Dislocation and Degradation of Endoplasmic Reticulum Proteins

M. Kikkert (✉) · G. Hassink · E. Wiertz

Department of Medical Microbiology, Leiden University Medical Center (LUMC), Albinusdreef 2, 2333 ZA Leiden, The Netherlands
m.kikkert@lumc.nl

Abstract Ubiquitination is essential for the dislocation and degradation of proteins from the endoplasmic reticulum (ER). How exactly this is regulated is unknown at present. This review provides an overview of ubiquitin-conjugating enzymes (E2s) and ubiquitin ligases (E3s) with a role in the degradation of ER proteins. Their structure and functions are described, as well as their mutual interactions. Substrate specificity and functional redundancy of E3 ligases are discussed, and other components of the ER degradation machinery that may associate with the ubiquitination system are reviewed.

Abbreviations

ALS	Amyothrophic lateral sclerosis
AMF	Autocrine motility factor
AMFR	Autocrine motility factor receptor
AR-JP	Autosomal-recessive juvenile parkinsonism
CFTR	Cystic fibrosis transmembrane conductance regulator
CHIP	Carboxyl terminus of Hsc70-interacting protein
CPY*	Mutated carboxypeptidase yscY
D2	Type 2 iodothyronine deindinase
E1	Ubiquitin-activating enzyme
E2	Ubiquitin-conjugating enzyme
E3	Ubiquitin ligase
ER	Endoplasmic reticulum
FBA	F-box-associated
Fbs	F-box recognizing sugars
HECT	Homologs to E6-AP C-terminus
HMGR	3-Hydroxy-3-methylglutaryl-coenzyme A reductase
IBR	In-between RING
LAP	Leukemia-associated protein
MARCH	Membrane-associated RING-CH
MHC	Major histocompatibility complex
Pael-R	Parkin-associated endothelin receptor-like receptor
PD	Parkinson's disease
PHD	Plant homeo domain
RA	Rheumatoid arthritis
RING	Really interesting new gene
SCF	Skp, cullin, F-box
SOD1	Superoxide dismutase 1
T3	3,5,3′-Triiodothyronine
TCR-α	α-Subunit of the T-cell receptor
TfR	Transferrin receptor
TNF-α	Tumor necrosis factor α
TPR	Tetratricopeptide
UBC	Ubiquitin-binding core
UBL	Ubiquitin-like
UPR	Unfolded protein response
VCP	Valosin-containing protein

1 Introduction

Since the degradation of many proteins from the ER has been found to involve their dislocation to the cytosol, deglycosylation by an N-glycanase, and destruction by the proteasome (Ward et al. 1995; Hampton et al. 1996; Biederer et al. 1996; Hiller et al. 1996; Wiertz et al. 1996a, 1996b; Sommer and Wolf 1997; Plemper et al. 1997), the obvious question arose about the involvement of ubiquitin in this process. It had long been known that cytosolic proteins destined for degradation by the proteasome are tagged with multiple copies of the 76 amino acid protein ubiquitin. At present, it is clear that degradation substrates from the ER are not only ubiquitinated in the cytosol for their recognition by the proteasome (Hirsch et al. 2004), but the ubiquitination machinery is also essential for the dislocation of these substrates from the ER to the cytosol (de Virgilio et al. 1998; Shamu et al. 2001; Kikkert et al. 2001). Here, we review the role of ubiquitination in dislocation and degradation of ER proteins, and in particular the identity and function of the ubiquitination enzymes that are specifically or aspecifically involved in this process. Besides E2 enzymes, especially the ubiquitin ligases, or E3s, take a key position in the ubiquitination of ER proteins. The unraveling of the processes at the ER membrane has only just begun, and many important discoveries will undoubtedly be made in the near future.

1.1 Ubiquitination: A Versatile Tool Within the Cell

The attachment of ubiquitin to a protein involves three steps (reviewed in Fang and Weissman 2004; Hirsch et al. 2004). The first step is the activation of ubiquitin by the ubiquitin activation enzyme, or E1, to prepare ubiquitin for further transfer. Ubiquitin-conjugating enzymes, or E2s, subsequently bind the activated ubiquitin via a conserved cysteine residue. It is thought that the E2 enzyme, probably in association with an E3 enzyme (see below), determines how the ubiquitin molecules will be attached to each other and the substrate. A single, monoubiquitin attachment, or a polyubiquitin chain linked through certain lysines in the ubiquitin molecules is thus formed on the target protein. A ubiquitin molecule can form an isopeptide linkage between its C-terminal glycine residue and one of the lysines at positions 6, 11, 29, 48, or 63 on the next ubiquitin molecule. Whereas K48 linkage is mostly involved in proteasomal degradation, K63-based ubiquitin linkage has been associated with other cellular functions (Fang and Weissman 2004). K11 and K29 have also been associated with protein degradation by the proteasome (Fang and Weissman 2004). The ubiquitin protein ligases, or E3 enzymes, accomplish the third step in the process. They bind the substrate and facilitate the actual transfer of one or more ubiquitin molecules from the E2 enzyme to a lysine or the N-terminus of the selected protein (Fang and Weissman 2004).

While ubiquitin was initially discovered as a tag for protein degradation in the cytosol, we now know that ubiquitination also plays an important part in many other cellular functions. DNA repair in the nucleus, regulation of translation, activation of transcription factors and kinases, and proteasomal degradation all require ubiquitination (Fang and Weissman 2004). Protein sorting and trafficking within the secretory system of the cell is also regulated by ubiquitination (Fang and Weissman 2004). In this context, (mono-) ubiquitination provides sorting signals for transport of protein cargo from the plasma membrane and from the trans-Golgi network to endosomes, and subsequently from endosomes to the multivesicular body and the lysosome (Hicke 1999; Strous and van Kerkhof 2002; Stahl and Barbieri 2002; Hicke and Dunn 2003; Marmor and Yarden 2004; Sigismund et al. 2004). Shuttling of transcription factors between the nucleus and the cytosol is also regulated by ubiquitin (Shcherbik and Haines 2004). The role of ubiquitin in degradation of ER proteins seems to be twofold. First, the sorting of substrates from the ER into the cytoplasm via dislocation fully depends on a functional ubiquitination machinery (de Virgilio et al. 1998; Shamu et al. 2001; Kikkert et al. 2001). Second, once the substrate has reached the cytosol, ubiquitination is needed for the protein to be recognized by the proteasome, as is the case for most genuine cytosolic substrates. Thus, ubiquitination has more than one key function in the degradation of ER proteins, and it is therefore a central component of this process.

Defective ubiquitination of ER proteins forms the basis of diseases such as autosomal-recessive juvenile parkinsonism (Takahashi and Imai 2003; Takahashi et al. 2003), type 2 diabetes mellitus (Allen et al. 2004), and rheumatoid arthritis (Amano et al. 2003), illustrating the crucial importance of the ubiquitin system in relation to the degradation of ER proteins.

1.2
The Role of Ubiquitin in the Dislocation of Endoplasmic Reticulum Proteins: A Hypothesis

The process of retrograde transport of ER proteins to the cytosol for their degradation depends on ubiquitination (de Virgilio et al. 1998; Shamu et al. 2001; Kikkert et al. 2001). This is illustrated by experiments in which ubiquitination is blocked by a temperature-sensitive mutation in the ubiquitin activating enzyme (E1). At the restrictive temperature, ER degradation substrates are retained in the ER membrane (de Virgilio et al. 1998; Yu and

Kopito 1999; Kikkert et al. 2001). It is reasonable to assume that the degradation substrates become polyubiquitinated themselves, as was indeed shown (Shamu et al. 1999). However, ubiquitination of substrates before their dislocation is difficult to envisage for ER-lumenal substrates, or proteins that lack lysines in their cytosolic domains. Still, ubiquitination is essential for the retrotranslocation of ER lumenal substrates such as CPY* (Biederer et al. 1997) and mutated ribophorin I (de Virgilio et al. 1998). Two observations may provide hints for an explanation of this apparent paradox. First, it has been suggested that the dislocation may be divided into two distinct steps (Elkabetz et al. 2004). The first step would be the relocation of ER lumenal proteins, or lumenal protein domains, to the cytosolic side of the ER membrane, while they remain associated with this membrane. The second step is the release of the degradation substrate from the cytosolic side of the ER membrane into the cytosol for degradation by the proteasome (Elkabetz et al. 2004). Substrates that were initially lumenal could thus be ubiquitinated while associated to the cytosolic side of the ER membrane. This then may be essential for the actual release into the cytosol, which is thought to be directed by the p97-Ufd1-Npl4 complex (Ye et al. 2001; Elkabetz et al. 2004). This view is supported by the observation of ER membrane-associated ubiquitinated intermediates (Shamu et al. 1999).

Alternatively, other observations could explain the membrane paradox. Both TCR-α and MHC class I heavy chains are well-known ER degradation substrates, and each contain a number of lysines. However, removal of the lysines from the cytosolic tail of MHC class I heavy chain does not influence its dislocation or degradation (Shamu et al. 1999). Moreover, removal of all of the lysines from TCR-α results in dislocation and proteasomal degradation with kinetics indistinguishable from that of wild type TCR-α (Yu et al. 1997). These data indicate that, although these proteins have lysines accessible to the ubiquitination machinery, these are not important for the removal of the proteins to the cytosol. Ubiquitination may still take place at the N-terminus of the lysineless TCR-α to facilitate its release from the ER membrane, which will require prior relocation to the cytosolic side of the ER membrane. However, a tempting hypothesis is that ubiquitination of the substrate is not required for its dislocation from the ER membrane. This then implicates that (an)other factor(s) may be ubiquitinated *in trans* to facilitate dislocation of ER proteins.

2
E2s and E3s Involved in Dislocation and Degradation of Endoplasmic Reticulum Proteins

The fact that ubiquitination is essential to the degradation of ER proteins is undoubted at present. However, the processes at the ER membrane that regulate the dislocation and degradation of proteins from the ER are far from understood. In this paragraph we will focus on ubiquitinating enzymes with a known function in the degradation of ER proteins.

2.1
Ubiquitin Conjugating Enzymes (E2s) Involved in Degradation of Endoplasmic Reticulum Proteins

Ubiquitin-conjugating enzymes are characterized by a core of 150 amino acids, the UBC domain, which includes a conserved cysteine to which the activated ubiquitin can bind. The UBC domain demonstrates around 35% identity among protein family members (Weissman 2001; Fang and Weissman 2004). In yeast, Ubc7p is ultimately connected to degradation of proteins from the ER, while Ubc6p and Ubc1p have also been associated with this process (Friedlander et al. 2000; Kiser et al. 2001; Bays et al. 2001; Botero et al. 2002). Ubc7p is recruited to the yeast ER membrane by Cue1p, a membrane-associated protein with a ubiquitin-binding domain known as the Cue-domain (Biederer et al. 1997; Ponting 2000). Ubc6p localizes to the ER by means of a C-terminal membrane anchor (Sommer and Jentsch 1993). Murine homologs of yeast Ubc6p and Ubc7p, MmUbc6 and MmUbc7, respectively, have been characterized by Tiwari and Weissman (Tiwari and Weissman 2001). Analogous to its yeast counterpart, MmUbc6 localizes to the ER through a membrane anchor. An apparent Cue1p homolog that functions as a recruiter of Ubc7 has not been found in vertebrates. However, the ER degradation-associated E3 ligase gp78, which is further described below, itself contains a Cue domain that strongly binds MmUbc7 (Tiwari and Weissman 2001). Some reports claim that degradation of particular ER substrates depends on either Ubc6 or Ubc7. MmUbc7 regulates the degradation of endogenous Inositol 1,4,5-triphosphate receptor, an ion channel-forming ER resident protein, whereas MmUbc6 is not able to do this (Webster et al. 2003). Others have shown that both Ubc6 and Ubc7 are instrumental to the degradation of a single substrate. An inactive form of MmUbc7 renders TCR-α in the membrane fraction, indicating that MmUbc7 plays a role in TCR-α degradation (Tiwari and Weissman 2001). However, others have shown that in mammalian cells, besides MmUbc7, MmUbc6 is also important for the degradation of TCR-α (Lenk et al. 2002). MmUbc6 and

MmUbc7 both influence the degradation of type 2 iodothyronine deiodinase (D2) that produces 3,5,3'-triiodothyronine (T3), which is essential for brain development (Botero et al. 2002; Kim et al. 2003). In summary, Ubc6 and Ubc7 are by far the E2 enzymes most frequently associated with degradation of ER proteins in yeast and mammalian cells. They are believed to be recruited by E3 enzymes, but how exactly this is regulated is unknown as yet.

2.2 Ubiquitin Ligases (E3s) Involved in Degradation of Endoplasmic Reticulum Proteins

Most of the E3 ligases that will be discussed below are not restricted to a function in degradation of ER proteins alone. Often they are capable of directing the degradation of both ER and cytoplasmic substrates. However, the two mammalian ER membrane-integrated ubiquitin ligases, HRD1 and gp78, and yeast Hrd1p, may be fully dedicated to the degradation of ER proteins, since no cytoplasmic substrates have been found for these enzymes yet. The number of E3 ubiquitin ligases identified is increasing rapidly at the moment. Three classes are recognized within this group of proteins, each characterized by a distinct motif that defines the E3 ligase function. The first class of E3 enzymes possess a HECT domain, a C-terminal region of approximately 350 residues, which was first recognized in E6-AP, an E3 enzyme encoded by human papillomavirus (Huibregtse et al. 1995). The conserved cysteine residue in this motif forms a covalent thiol-ester intermediate with ubiquitin, which is subsequently transferred to the substrate (Huibregtse et al. 1995; Fang and Weissman 2004). The second and third groups contain a RING-finger or a U-box motif, respectively. RING-finger proteins were identified long before an association with ubiquitin ligase activity was recognized, but at present, many proteins containing a RING-finger have been found to possess this activity *in vitro* (Lorick et al. 1999; Weissman 2001; Fang and Weissman 2004). The RING motif consists of a series of eight conserved cysteines and histidines, which bind two zinc atoms and form a structure of "cross-braced" rings. The middle two residues in the motif comprise either one or two histidines, resulting in three subclasses of RING-finger motifs: classical or RING-HC, RING-CH, and RING-H2. U-box domains have the same fold as RING-fingers, but contain hydrogen bonds instead of zinc ions (Aravind and Koonin 2000). Unlike HECT domain ubiquitin ligases, RING- and U-box-containing E3 enzymes are thought not to form intermediates with ubiquitin. The exact mode of transfer of ubiquitin to the substrate facilitated by E3 ubiquitin ligases is unknown to date (Weissman 2001; Fang and Weissman 2004). Table 1 shows an overview of the E3 ligases discussed in this review.

Table 1 Overview of ubiquitin ligases with a role in degradation of ER proteins

E3 ligase	Section	Functional domains	Localization, topology, size (mol. weight on gel)	(Model) ER substrate(s)	Associated disease(s)	Remarks
S. cerevisiae						
Hrd1p/ Der3p	2.2.1.1	RING-H2	ER membrane, 6 TM, RING cytosolic, 551 aa	HMGR, CPY*, sec61–2p, CFTR (ectopically expressed)	–	Human homologs: HRD1 and AMFR/gp78
Doa10p	2.2.2.1	RING-CH	ER/nuclear membrane, 10–14 TM, 1319 aa (151 kDa)	Ubc6, Pma1-D378N, Ste6–166	–	Human homolog: TEB4
Rsp5p	2.2.3.1	HECT, C2, WW	Cytosol, nucleus, 809 aa	CPY*, sec61–2p, both only under conditions of ER stress	–	Part of the HIP pathway
H. sapiens						
HRD1	2.2.1.3	RING-H2	ER membrane, 6 TM, 616 aa (82 kDa)	CD3-δ, TCR-α, insulin	Rheumatoid arthritis	Homolog of Hrd1p/Der3p
AMFR/ gp78	2.2.1.2	RING-H2, leucine zipper	ER membrane, 7 TM, 643 aa (78 kDa)	CD3-δ, apolipo-protein B100	Metastasis marker for several cancers	Homolog of Hrd1p/Der3p
TEB4	2.2.2.2	RING-CH	ER membrane, 13 TM, 910 aa (97 kDa)	–	–	No substrates identified yet, homolog of Doa10p

Table 1 (continued)

E3 ligase	Section	Functional domains	Localization, topology, size (mol. weight on gel)	(Model) ER substrate(s)	Associated disease(s)	Remarks
SCF^{Fbs1}	2.2.3.2	F-box, FBA	Cytosol, membrane-associated, 296 aa (42 kDa)	Integrinβ1, CFTR-GFP, TCR-α	–	Targets N-linked glycans on dislocated ER proteins
SCF^{Fbs2}	2.2.3.2	F-box, FBA	Cytosol, membrane-associated, 293 aa	TCR-α	–	Targets N-linked glycans on dislocated ER proteins
$SCF^{\beta\text{-TrCP}}$	2.2.3.2	F-box, WD	Cytosol, 605 aa (64 kDa)	CD4, but only when HIV-encoded Vpu binds to β-TrCP	AIDS	–
CHIP	2.2.3.3	U-box, TPR	Cytosol, 303 aa	CFTR, Pael-R (latter via Parkin)	Cystic fibrosis, Parkinson's, Alzheimer's	Targets Hsp/c 70/90 chaperone substrates
Parkin	2.2.3.4	RING-IBR-RING, UBL	ER-associated, but no TM regions, 465 aa	Pael-R	Autosomal recessive juvenile parkinsonism	–
Malin	2.2.3.5	RING-HC, NHL-repeats	ER, nucleus, 395 aa	–	Lafora disease	Function in ER degradation has to be confirmed

2.2.1
The HRD1-Like Family of Ubiquitin Ligases

Hrd1p of *Saccharomyces cerevisiae* was the first E3 ubiquitin ligase identified that contributes to the degradation of ER proteins (Bays et al. 2001; Deak and Wolf 2001). In mammalians, two homologs have been identified: autocrine motility factor receptor (AMFR) or gp78 (Fang et al. 2001) and HRD1 (Kaneko et al. 2002; Nadav et al. 2003; Kikkert et al. 2004).

2.2.1.1
Saccharomyces Cerevisiae Hrd1p

Hrd1p/Der3p of *S. cerevisiae* was identified independently by Hampton and co-workers (Hampton et al. 1996) and Wolf and co-workers (Bordallo et al. 1998). Hrd1p/Der3p functions as an E3 ubiquitin ligase in the degradation of *S. cerevisiae* Hmg2p, one of the yeast isozymes of 3-hydroxy-3-methylglutaryl-coenzyme A reductase (HMGR) (Hampton et al. 1996; Bays et al. 2001). HMGR is the rate-limiting enzyme in the mevalonate pathway, in which sterols and a myriad of essential isoprenoids are synthesized. The levels of HMGR are regulated by its degradation, depending on the cellular demands for mevalonate-derived sterols and nonsterol metabolites (Hampton 2002). Hrd1p/Der3p is also involved in degradation of other ER proteins, including CPY* and sec61-2p (Bordallo et al. 1998; Bordallo and Wolf 1999). Hrd1p/Der3p is a multispanning membrane protein encompassing six transmembrane domains, and a C-terminal RING-H2 domain located in the cytosol (Deak and Wolf 2001). In yeast, Hrd1p is found in a 1:1 complex with Hrd3p, a lumen-oriented ER membrane protein that stabilizes Hrd1p and modulates its ligase activity (Gardner et al. 2000). Hrd3p is therefore essential for the proper functioning of Hrd1p in yeast. Hrd1p acts with the E2 enzymes Ubc1p and Ubc7p. Ubc7p seems to be more important in the context of ER protein dislocation and degradation (Bays et al. 2001). Initially, it was hypothesized that Hrd1p may be the central ubiquitin ligase that served the degradation of all yeast ER proteins together with Hrd3p, Der1p, Ubc1p, and Ubc7p (Gardner et al. 2001). However, while a number of substrates depend on Hrd1p for their degradation from the ER, a number of others do not (Wilhovsky et al. 2000; Hill and Cooper 2000). Besides Hrd1p, two other E3 ligases with a role in the degradation of ER substrates have been identified in yeast to date, Rsp5p and Doa10 (see below). In mammalians, two homologs of yeast Hrd1p have been identified, AMFR/gp78 and HRD1, which will be discussed in the next sections.

2.2.1.2
Gp78/AMFR

AMFR was identified as a 78-kDa cell surface receptor for autocrine motility factor (AMF). Binding of AMF to AMFR stimulates random and directed cell motility, accounting for the contribution of AMFR in metastasis (Nabi et al. 1990, 1992; Watanabe et al. 1991). Gp78/AMFR is a known marker for metastasis and has been associated with bad prognoses in various cancers such as bladder cancer, colorectal cancer, esophageal cancer, and gastric cancer (Nakamori et al. 1994; Maruyama et al. 1995; Hirono et al. 1996; Otto et al. 1997). When overexpressed in NIH3T3 mouse fibroblasts, gp78/AMFR causes altered morphology and growth characteristics of these cells pointing toward a transformation phenotype. In nude mice, gp78 overexpression induces tumors (Onishi et al. 2003). It was noted that gp78/AMFR contains a RING-H2 domain, as well as a sequence patch denoted as a Cue domain (Ponting 2000), downstream of a region containing 7 predicted N-terminal transmembrane domains and a leucine zipper domain (Shimizu et al. 1999). Additionally, human gp78 shows limited homology to yeast Hrd1p, together prompting the idea that gp78 could be a ubiquitin ligase with a role analogous to Hrd1p in yeast (Ponting 2000). Indeed, it was shown that gp78 has RING-finger-dependent ubiquitin ligase activity, and seems to be largely localized to the ER (Fang et al. 2001). It can expedite the degradation of a well-known ER substrate, CD3-δ, an activity which is dependent on the integrity of the gp78 RING-finger, and the presence of the gp78 transmembrane region (Fang et al. 2001). Another known ER substrate, Apolipoprotein B100, is also degraded with the help of gp78 (Liang et al. 2003). The degradation of Gp78 itself is proteasome and Ubc7-dependent, and mediated by autoubiquitination dependent on its own RING-finger (Fang et al. 2001). Interestingly, Gp78 binds to p97/VCP-Ufd1-Nlp4 (Zhong et al. 2004), an observation that further establishes the role of gp78 in the degradation of ER proteins.

The Cue domain, present in gp78, has earlier been recognized in the yeast protein Cue1p, which has a role in recruiting yeast Ubc7p to the ER membrane (Biederer et al. 1997; Ponting 2000). Indeed, gp78 interacts with mammalian (murine) Ubc7, but surprisingly, the Cue domain is not sufficient for this binding (Fang et al. 2001). Instead, the Cue domain together with the remaining part of the C-terminus of gp78 binds mouse Ubc7 efficiently (Fang et al. 2001). Recently, it has been found that the Cue domain binds monoubiquitin (Shih et al. 2003). Vsp9, a yeast protein involved in the endocytic pathway, contains a Cue domain that binds to monoubiquitin and this mediates the intramolecular monoubiquitination of Vsp9 elsewhere on the protein. The affinity for monoubiquitin differs among different Cue-domain proteins, but seems to depend on three amino acids preceding the motif. The Gp78 Cue domain is in a context that favors the binding of ubiquitin (Shih et al. 2003). How the binding of E2 and/or monoubiquitin to the Cue domain and Cue domain mediated auto-mono-ubiquitination relates to the functions and characteristics of gp78 remains to be elucidated.

2.2.1.3 Human HRD1

Besides gp78/AMFR, another, closer, homolog of yeast Hrd1p, HRD1, has been identified (Kaneko et al. 2002; Nadav et al. 2003; Kikkert et al. 2004). HRD1 is characterized by six predicted transmembrane regions and a cytosolic RING-H2 finger (Kikkert et al. 2004), thereby strongly resembling yeast Hrd1p (Bays et al. 2001; Deak and Wolf 2001). *In vitro* assays have shown that human HRD1 has E3 ubiquitin ligase activity that is specific for K48 linkage on ubiquitin (Nadav et al. 2003; Kikkert et al. 2004). UBC7 is able to cooperate with HRD1 (Kikkert et al. 2004), but no binding of MmUbc7 (or MmUbc6) to HRD1 could be detected (M. Kikkert, unpublished observations). Intracellular localization studies indicate that HRD1 is an ER-resident protein (Kaneko et al. 2002; Nadav et al. 2003; Kikkert et al. 2004). The mRNA encoding HRD1 was found in most tissues analyzed, but is particularly abundant in pancreas, liver and skeletal muscle tissues (Kaneko et al. 2002; Nadav et al. 2003; M. Kikkert/T. van Laar, unpublished observations). Close homologs of human HRD1 are found in other mammalians such as *Mus musculus* (95% identity with human HRD1) and *Rattus norvegicus*, in *Drosophila melanogaster*, and even in *Arabidopsis thaliana*, indicating that HRD1 is highly conserved among eukaryotes. The degradation of model substrates CD3-δ and TCR-α is facilitated by HRD1, indicating that HRD1 has a role in the degradation of ER proteins (Kikkert et al. 2004). The sterol-regulated degradation of HMGR, however, does not seem to be directed by HRD1, in contrast to the basal turnover of HMGR that is independent of sterol regulation (Kikkert et al. 2004). HRD1 is upregulated upon ER stress in an Ire1p-dependent way (Kaneko et al. 2002), suggesting that HRD1 is part of the unfolded protein response (UPR) effector protein-set. Experimental hypoxia-ischemia in mouse brains also causes elevated levels of HRD1, probably due to ER dysfunction (Qi et al. 2004). The (over-)expression of HRD1 protects cells from ER stress-induced apoptosis (Kaneko et al. 2002), supposedly because of its function in the degradation of ER proteins.

Several reports point to a role of HRD1 in common human diseases. The strongest indications for the connection of HRD1 with disease are found in rheumatoid arthritis (RA). Amano and colleagues (Amano et al. 2003) have shown that mice that constitutively overexpress HRD1 develop arthritis spontaneously in a significant number of cases. Additionally, down-modulation of HRD1 expression by RNAi protects mice against collagen-induced arthritis. Increased levels of HRD1 expression are found in joints of human RA patients, and this seems to inhibit apoptosis in the synovial cells of the joints. Different forms of stress in the joint induce proliferation of synovial cells, and apoptosis is needed to reduce the amount of synoviocytes after the stress has faded. If the reduction of synovial cells does not take place, because increased levels of HRD1 protect the cells from dying, this is thought to cause inflammation characteristic of RA (Amano et al. 2003). Whether normal apoptosis of synoviocytes is indeed ER stress-induced has to be determined.

Recently, a role for ER stress in the development of type 2 diabetes has been confirmed (Ozcan et al. 2004; Allen et al. 2004; Nakatani et al. 2005). The Akita mouse, a well-established model for type 2 diabetes, harbors a heterozygous mutation of the insulin-2 gene. The mutant insulin protein causes stress in the ER because of its misfolding, and this results in the upregulation of UPR genes including HRD1 (Allen et al. 2004). The mutant insulin-2 protein seems to be a substrate for HRD1 in Akita mouse cells, raising the possibility of HRD1-based therapies for type 2 diabetes (Allen et al. 2004).

A third human disease that may be connected to the functions of HRD1 in degradation of ER proteins and ER stress relief is autosomal recessive juvenile parkinsonism (AR-JP). Earlier, the accumulation of Pael-R in the ER has been recognized as a hallmark of this disease. Pael-R accumulation is caused by mutation of the *parkin* gene, encoding an E3 ligase that facilitates degradation of Pael-R (see below). Recently, it has been shown that HRD1 is also able to reduce the levels of Pael-R by degradation from the ER (Kaneko and Nomura 2004) . This suggests that upregulation of HRD1 can relieve the ER stress caused by Parkin mutations, but apparently the UPR-induced HRD1 upregulation that supposedly takes place upon Pael-R accumulation is not sufficient to prevent the disease.

2.2.2 Doa 10p and TEB4

S. cerevisiae Doa10p (Swanson et al. 2001) and mammalian TEB4 (Hassink et al. 2005) are E3 ubiquitin ligases that contain a RING-finger of unusual configuration. Both are associated with degradation of ER proteins.

2.2.2.1 *Saccharomyces Cerevisiae* Doa10p

In a search for genes responsible for the degradation of the cytosolic yeast mating factor-α2, Doa10p was identified as a novel *S. cerevisiae* E3 ubiquitin ligase (Swanson et al. 2001). Doa10p is a multimembrane-spanning RING-finger containing ubiquitin ligase that resides in the ER and the nuclear envelope (Swanson et al. 2001). It promotes ubiquitination of proteins with a degradation signal denoted *Deg1,* which is also present within the N-terminal 62 residues of α2. Doa10 collaborates with the E2 enzymes Ubc6 and Ubc7 (Swanson et al. 2001).

Doa10 harbors the unusual RING-CH configuration (Swanson et al. 2001). Proteins containing this RING-CH motif have earlier been associated with transcriptional regulation and DNA binding (Aasland et al. 1995; Saha et al. 1995; Lyngso et al. 2000; Kosarev et al. 2002), and designated as PHD-(plant homeo domain) (Schindler et al. 1993) or LAP-(leukemia-associated protein) domain containing proteins (Saha et al. 1995). These proteins do not function as E3 ubiquitin ligases. Aravind and colleagues (Aravind et al. 2003), however, were able to discern structural differences apart from cystein and histidine composition that made it possible to discriminate between RING-HC-containing proteins that act as ubiquitin ligases and PHD/LAP domain-containing proteins with other functions. This refinement placed Doa10 in the family of E3 ligases and not in the PHD/LAP domain-containing group of proteins.

Since Doa10 is localized in the ER membrane, it was sensible to test whether Doa10 is involved in degradation of ER proteins. Indeed, the degradation of the short-lived ER-resident E2 enzyme Ubc6 was markedly inhibited in a *Doa10Δ* yeast mutant (Swanson et al. 2001). Degradation of Pma1-D378N and Ste6-166, both misfolded forms of yeast plasma-membrane proteins, takes place from the ER and also depends on Doa10p (Wang and Chang 2003; Vashist and Ng 2004). It was found that degradation of either of these proteins does not depend on Hrd1p. The degradation of CPY*, which has been shown to depend on Hrd1p, was not influenced by Doa10p (Walter et al. 2001; Wang and Chang 2003; Vashist and Ng 2004). These results suggest that Hrd1p and Doa10 cooperate in yeast ER protein degradation, each serving a distinct subset of ER substrates. However, when human CFTR was ectopically expressed in yeast, its degradation depended on both Hrd1p and Doa10p. This was illustrated by the strong effect of deleting both E3s, whereas deleting either of them separately gave only modest effects on the degradation of CFTR (Gnann et al. 2004). These data suggest that Hrd1p and Doa10p are able to complement each other in the degradation of a single substrate. When Hrd1p and Doa10p are both deleted, yeast cells become extremely sensitive to ER stress, as well as to cadmium treatment, whereas deletion of only one of the two genes has only modest effects (Swanson et al. 2001). A temperature-sensitive mutation in Npl4p caused malfunctioning of the Cdc48p-Npl4p-Ufd1p complex (see "Bar-Nun" and below). The accumulation of ubiquitinated proteins in the (ER-) membrane that was caused by this Npl4p mutation could be suppressed by deleting both Doa10p and Hrd1p (Hitchcock et al. 2003). Together, these findings further illustrate that both proteins have a complementary role in the degradation of ER proteins and the neutralization of ER stress in yeast.

2.2.2.2
TEB4

We have recently identified TEB4 as the mammalian homolog of yeast Doa10p (Hassink et al. 2005). It was originally characterized as a transcript of the Cri-du-chat critical region on chromosome 5 and appears to be well conserved, as genes with a high degree of homology to TEB4 occur in many species. TEB4 contains 13 predicted transmembrane regions and has a RING-CH domain near its N-terminus. It exhibits UBC7-dependent E3 ligase activity *in vitro*, which is also ubiquitin lysine 48-specific (Hassink et al. 2005). While it promotes its own degradation in a RING-finger and proteasome-dependent fashion (Hassink et al. 2005), other substrates for TEB4 have not been found as yet. We tested the effect of overexpression of TEB4 and its RING-finger mutant on US11-dependent dislocation of MHC class I molecules, and on the degradation of UBC6. No effect was observed on the degradation of either of these substrates (Hassink et al., unpublished observations).

The putative role for TEB4 in ER protein degradation is, however, supported by its homology with *S. cerevisiae* Doa10p (Swanson et al. 2001), its ER localization, and the large number of transmembrane regions, the involvement of lysine 48 of ubiquitin in the E3 ligase activity, *in vitro* dependence on UBC7, and its (auto-) degradation by the proteasome (Hassink et al. 2005).

Membrane associated RING-CH (MARCH) proteins (Bartee et al. 2004), such as murine gamma herpesvirus 68 mK3 and Kaposi sarcoma herpesvirus encoded kK3 and kK5, inhibit the expression of MHC class I complexes and the co-stimulatory molecules ICAM-1 and B7.2 on the cell surface (Coscoy and Ganem 2000; Ishido et al. 2000a, 2000b; Coscoy et al. 2001; Coscoy and Ganem 2001; Hewitt et al. 2002; Lorenzo et al. 2002; Bartee et al. 2004). These E3 ligases mediate ubiquitin-dependent internalization of receptor molecules and their degradation in an endolysosomal compartment. Neither TEB4 nor its RING-finger mutant affected surface expression of immunomodulatory molecules such as MHC class I, Fas, TfR, CD4, and B7.2 (Hassink et al., unpublished observations), suggesting that TEB4 does not share this function with the other MARCH proteins.

2.2.3
Other Yeast and Mammalian Ubiquitin Ligases That Have a Role in the Degradation of Endoplasmic Reticulum Proteins

2.2.3.1
Saccharomyces cerevisiae Rsp5p

Rsp5p is a HECT domain-containing *S. cerevisiae* ubiquitin ligase (Yashiroda et al. 1996). It has an N-terminal C2 domain that binds membrane phospholipids in a calcium-regulated manner, and three WW domains, which are evolutionary conserved protein interaction modules of about 40 amino acids each that bind proline-rich ligands (Wang et al. 1999). The functions of Rsp5p in yeast are diverse, and deletion of the Rsp5 encoding gene is lethal (Hein et al. 1995; Hoshikawa et al. 2003). This E3 enzyme has been implicated in the sorting of plasma membrane proteins into multivesicular bodies and the vacuole (Katzmann et al. 2004). It ubiquitinates plasma membrane permeases such as Gap1, thereby initiating their endocytoses and vacuolar degradation (Hein et al. 1995; Galan et al. 1996). Rsp5 also ubiquitinates transcriptional regulators (Hoppe et al. 2000) and RNA polymerase II, and it therefore seems to have a general role in gene expression (Chang et al. 2000; Rodriguez et al. 2003). When CPY* or sec61-2p, both established yeast ER degradation substrates, are overexpressed, the Hrd1p-based pathway of degradation becomes overloaded. The subsequent unfolded protein response will result in the up-regulation of the Hrd1p pathway components. Additionally, however, an alternative pathway of degradation in which Rsp5p acts as a ubiquitin ligase is activated (Haynes et al. 2002). In this pathway, designated "Hrd1p-independent proteolysis" (HIP), Ubc4p and Ubc5p are used as E2s. Proteins involved in the transport from the ER to the Golgi, i.e., erv29p, play an essential role in this pathway. A functional unfolded protein response, based on Ire1p activation, is necessary for the induction of the HIP pathway (Haynes et al. 2002). The deletion of both the Hrd1p and the Rsp5p-based pathways completely stabilizes CPY* and sec61-2p, whereas deletion of either pathway results in less efficient stabilization (Haynes et al. 2002). Haynes and colleagues hypothesize that overloading of the ER may result in a spillover of degradation substrates into the Golgi. The Rsp5p-mediated degradation then forms an alternative route for these escaped proteins.

Rsp5p is the yeast ortholog of the human HECT E3 ligase Nedd4, for which no role in the degradation of ER proteins has been established so far. However, the functions of Nedd4 resemble those of Rsp5p in yeast (Fang and Weissman 2004). The question therefore remains whether an alternative degradation route induced by the UPR also exists in mammalian cells, resembling the

HIP pathway. If this is the case, Nedd4, or other members of the family of Nedd4-like ubiquitin ligases, may have a role in this pathway.

2.2.3.2 Mammalian SCF Ubiquitin Ligase Complexes

The multisubunit SCF ubiquitin ligases consist of four subunits, among which a small RING-H2 finger protein, Rbx1 (also named Roc1 or Hrt1), which is present in all SCF complexes and is responsible for the actual ubiquitin transfer. The E2 Ubc3/Cdc34 associates with Rbx in the SCF ubiquitin ligase. A member of the cullin protein family, i.e., Cullin1/Cdc58, organizes the complex by interacting with Rbx, an adaptor (Skp1) and an F-box protein (Deshaies 1999; Zheng et al. 2002). F-box proteins form a large family of proteins containing a mostly N-terminal 42–48 amino acid F-box motif that binds to the Skp1 adaptor of the complex. WD40 repeats or leucine-rich repeats (LRRs) in the carboxytermini of F-box proteins interact with phosphorylated substrates (Winston et al. 1999; Kipreos and Pagano 2000). SCF ubiquitin ligase complexes are well-studied and have functions in several important cellular processes such as cell-cycle phase transitions and regulation of transcription (Petroski and Deshaies 2005).

Yoshida and co-workers found that a particular SCF complex, consisting of Skp1, Cullin1, Rbx, and Fbx2, has a role in degradation of ER proteins (Yoshida et al. 2002). Fbx2 is a 42-kDa protein expressed in neurons that preferentially binds to proteins containing N-linked glycans with a diacetylchitobiose structure and mannose residues. It was shown that Fbx2 bound to mouse pre-integrinβ1, which contains high-mannose oligosaccharides, and not to deglycosylated integrinβ1. Integrinβ1 becomes an ER degradation substrate when it is expressed in excess over its binding partner intergrin-α in the ER. Glycosylated CFTR-GFP and TCR-α, two well-known ER degradation substrates, were also found to bind to Fbx2. A mutant lacking the F-box motif inhibited degradation of these ERAD substrates in a dominant negative fashion (Yoshida et al. 2002). The binding of Fbx2 occurred to glycosylated substrates in the cytosol and not in membranes, indicating that the SCF-Fbx2 E3 enzyme targets ER substrates that have been dislocated to the cytosol but have not been deglycosylated (Yoshida et al. 2002).

Other SCF complexes recognize phosphorylated (Hershko and Ciechanover 1998; Deshaies 1999; Kipreos and Pagano 2000) or hydroxylated (Ivan et al. 2001; Jaakkola et al. 2001) substrates, but apparently glycosylation is also a modification that can be targeted by SCF ubiquitin ligases. While Fbx2 is present only in the brain, another high-mannose oligosaccharide-binding F-box protein, Fbx6b, is expressed in almost all mouse tissues tested (Yoshida et al. 2003). The carbohydrate-binding F-box proteins were renamed into Fbs1 (F-box protein that recognizes sugar chains) for Fbx2, and Fbs2 for Fbx6b, respectively. Both Fbs1 and Fbs2 contain C-terminal F-box-associated domains (FBA domains) with which they interact with glycans, instead of the more common WD or LRR domains that are present on other, phosphorylation specific, F-box proteins. However, Fbx17 and FBG3 that both have an FBA domain do not bind glycoproteins (Yoshida et al. 2003). Fbs1 and Fbs2 seem to have slightly different specificities toward N-linked glycans, but both Fbs1 and Fbs2 efficiently bind $Man_{3-9}GlcNac_2$, a structure found abundantly on proteins present in the ER. In contrast to other lectins such as calnexin and calreticulin, which recognize terminal glucose residues, Fbs1 recognizes the inner chitobiose portion of this structure (Mizushima et al. 2004). This may implicate that Fbs1 and Fbs2 specifically target unfolded versions of glycoproteins, in which the chitobiose core is more accessible. Like Fbs1, Fbs2 is able to direct ubiquitination of TCR-α, suggesting a role in the degradation of ER proteins for these F-box proteins (Yoshida et al. 2002, 2003). Since both F-box proteins can also bind other types of N-glycans, it cannot be excluded that they are able to target glycosylated surface proteins that have been internalized and have somehow reached the cytosol.

β-TrCp is another F-box protein that is part of an SCF complex also containing Skp1 and Cullin1. When phosphorylated Vpu of human immunodeficiency virus (HIV) binds to the β-TrCP complex, it degrades CD4 from the ER (Margottin et al. 1998). By binding Vpu, the normal function of β-TrCP in TNF-α-induced degradation of IκB-α is inhibited (Bour et al. 2001). Besides IκB-α, β-TrCP normally degrades a number of other cytosolic substrates such as β-catenin, (Hart et al. 1999; Liu et al. 1999), Cdc25A (Busino et al. 2003), and Emi1 (Peters 2003; Guardavaccaro et al. 2003; Margottin-Goguet et al. 2003). Thus, it is only through the interference of the viral protein Vpu that β-TrCP acts as an E3 ligase in the degradation of CD4 from the ER, while degradation of cytosolic proteins is normally facilitated by this enzyme.

2.2.3.3 CHIP

The U-box-containing ubiquitin ligase CHIP (carboxyl terminus of Hsc70-interacting protein) is a representative of a small group of E3 enzymes that interact with protein-folding chaperones in the cytosol (Hatakeyama et al. 2004b). When CHIP binds a folding chaperone bound to a substrate, the complex is modulated from a folding machine into a degradation machine, resulting in the ubiquitination and degradation of the substrate (McDonough and Patterson 2003). Besides a U-box, the cytosolic ubiquitin ligase CHIP contains a tetratricopeptide (TPR) domain, with which it binds Hsp70/Hsc70 and Hsp90 (Ballinger et al. 1999). CHIP interacts functionally with the stress-responsive UBCH5 E2 enzyme, but not with UbcH7, and has in vitro E3 ligase activity that depends on its U-box domain (Jiang et al. 2001).

The binding between the proteasome on the one hand and the complex of Hsp70, the substrate, and CHIP on the other, is mediated by BAG-1 (Alberti et al. 2002). This co-chaperone contains a ubiquitin-like (UBL) domain, with which it binds to the proteasome (Luders et al. 2000). BAG-1 accepts substrates from Hsp70, and presents them to CHIP while CHIP binds directly to BAG-1 (Demand et al. 2001). BAG-1 can itself be ubiquitinated by CHIP with E2s of the Ubc4/5 family, whereby lysine-11 linked ubiquitin chains are formed. This does not induce degradation of BAG-1, but instead promotes association of BAG-1 with the proteasome (Alberti et al. 2002).

A number of substrates whose degradation is facilitated by CHIP have been identified, and again, this E3 enzyme can serve both cytosolic and ER-associated substrates. Among the cytosolic substrates of CHIP is a protein called tau, a microtubule-binding protein whose accumulation in fibrils is implicated in so-called tauopathies, such as Parkinson's and Alzheimer's diseases (Hatakeyama et al. 2004a). Another example is the cytosolic glucocorticoid receptor, a cytosolic Hsp90 substrate, which is targeted for degradation by the proteasome when ubiquitinated by CHIP (Connell et al. 2001). CHIP can ubiquitinate Hsc70 that binds to mutant superoxide dismutase 1 (SOD1), but not mutant SOD1 itself. The ubiquitination of Hsc70 mediates the degradation of SOD1 mutants in an indirect way (Urushitani et al. 2004). Mutations in SOD1 are linked to familial amyotrophic lateral sclerosis (ALS) (Bruijn et al. 2004). The E2A transcription factor E47 is also degraded with the help of CHIP. CHIP acts as a carrier protein in this case, stimulating the binding of E47 to the Skp2 (F-box) component of SCF ubiquitin ligase. This SCF ligase subsequently ubiquitinates E47 in a phosphorylation-dependent manner (Huang et al. 2004).

CHIP is involved in the degradation of two ER proteins. Cystic fibrosis transmembrane conductance regulator (CFTR) is an ER-associated protein with 12 transmembrane domains that form an ion channel (Harris and Argent 1993). Mutations in this protein result in cystic fibrosis, which is the most common fatal inherited disease in Caucasian populations (Harris and Argent 1993). Folding of the cytosolic domain of CFTR is chaperoned by Hsp70, and while binding Hsp70, CFTR can be targeted for degradation from the ER by CHIP (Meacham et al. 2001). Indirectly, CHIP is also involved in the degradation of Parkin associated endothelian receptor-like receptor (Pael-R) from the ER. Pael-R is a substrate of the RING-containing ubiquitin ligase Parkin (see next section). CHIP is up-regulated upon accumulation of unfolded Pael-R, and binds to the complex of Hsp70, Parkin, and Pael-R (Imai et al. 2002). CHIP binding then enhances ubiquitination of Pael-R by Parkin (Imai et al. 2002). In Sect. 2.2.3.4, the connection between Parkin and CHIP will be discussed in more detail.

2.2.3.4 Parkin

Parkin is among the best studied E3 ligases at present. This is undoubtedly related to the connection of Parkin dysfunctionality to autosomal recessive juvenile parkinsonism (AR-JP), an inherited form of Parkinson's disease (PD) (Kitada et al. 1998). The Parkin gene is expressed primarily in neurons, with a developmentally regulated pattern of expression in the brain (Kitada et al. 2000; Wang et al. 2001), which is in agreement with its relation to Parkinson's disease.

Parkin has two RING fingers with an in-between sequence (IBR, in-between RING), and an amino-terminal ubiquitin-like domain (UBL) with which it can bind Rpn10, a 19S cap component of the proteasome (Sakata et al. 2003; Tsai et al. 2003). Mutation of the UBL domain of Parkin is connected to Parkinson's disease, indicating that this domain is important for its function (Tsai et al. 2003).

Parkin associates and acts with the UBC6 and UBC7 ubiquitin conjugating enzymes (Imai et al. 2001). Like HRD1, Parkin protects against ER stress-induced apoptosis. However, when Parkin is knocked out in mice, substrates do not accumulate (Goldberg et al. 2003; Palacino et al. 2004; Lorenzetti et al. 2004), suggesting redundant E3 activity. Parkin is localized at the ER membrane, as shown in light microscopy and electron microscopy analyses (Imai et al. 2002). However, it does not have transmembrane domains, and therefore has to be recruited to the ER by an (unknown) receptor, or by the Hsp70 machinery (see below). Although a number of substrates for Parkin have been identified, the exact mechanism of dopaminergic neuron loss as seen in Parkinson's disease has not yet been clarified. Among the substrates for Parkin that have been found to date are CDCrel-1, a septin family member present in presynaptic axon terminals of inhibitory neurons (Zhang et al. 2000), the p38 subunit of amino-acyl tRNA synthetase (Corti et al. 2003), and synaptotagmin XI, a protein involved in maintaining synaptic function (Huynh et al. 2003; Glass et al. 2004). CyclinE is degraded under the influence of a multisubunit SCF ubiquitin ligase, composed of Parkin, Cullin1, Skp1, and the F-box protein hSEL-10/Fbw7 (Staropoli et al. 2003). In this way, Parkin is directly involved in the regulation of the cell cycle, and it is therefore not surprising that mutations in the Parkin gene are also linked to breast and ovarian cancers, as well as parkinsonism (Cesari et al. 2003; Denison et al. 2003). Another substrate of Parkin is synphilin-1, which binds to α-synuclein (Chung et al. 2001). An O-glycosylated form (22 kDa) of α-synuclein is also a substrate of Parkin (Shimura et al. 2001). Alpha-synuclein is the main constituent of Lewy bodies (Chung et al. 2001), which are the cellular aggregates associated with Parkinson's disease (McNaught and Olanow 2003).

All substrates described are cytosolic, and some of them are not associated with Parkinson's disease. The only membrane-bound protein that has been recognized as a Parkin substrate is Pael-R, a G-protein coupled receptor protein (Imai et al. 2001). As a result of mutations in Parkin, Pael-R accumulates in the ER, which causes ER stress. This condition has been associated with AR-JP (Takahashi and Imai 2003). Pael-R is also found in Lewy bodies (Murakami et al. 2004).

While Parkin has many cytosolic substrates, it is up-regulated upon ER stress, but not by other forms of cellular stress. Parkin protects cells against ER stress-induced apoptosis through its RING-finger-dependent E3 ligase activity (Imai et al. 2000).

The function of Parkin is regulated in a number of ways. Like many other E3 ligases, Parkin can ubiquitinate itself to reduce its own levels (Zhang et al. 2000). Deletion of its UBL domain increases Parkin's expression levels, suggesting that this domain also has a role in the regulation of Parkin levels (Finney et al. 2003). The ER substrate of Parkin, Pael-R, is assisted in entering into the ER by the chaperones Hsp70 and Hdj-2. The binding of Hsp70 to Pael-R may prevent the ubiquitination of Pael-R by Parkin. When unfolded Pael-R is generated in the ER, it is retrotransported into the cytosol. Hsp70 and Hdj-2 then transiently bind to prevent the unfolded Pael-R from becoming insoluble. CHIP is up-regulated upon the misfolding of Pael-R and binds to Hsp70, thereby promoting the release of Hsp70 and Hdj-2 from Pael-R. CHIP also associates and cooperates with Parkin and E2s on the ER surface, to promote ubiquitination of Pael-R (Imai et al. 2002).

Parkin is regulated at the transcription level by the N-myc transcription factor, through binding of this factor to a transcription regulation motif in the *parkin* promotor, which resembles a so-called E-box motif that is conserved across several species. N-myc transcription factors are critically involved in neuronal development and tumorigenesis. N-myc also seems to regulate transcription of other PD-associated genes (West et al. 2004).

Upon incubation with proteasome inhibitors, Parkin is relocalized into aggregates in the centrosome area, where ubiquitin, proteasomes, and some of the Parkin substrates are also found. Parkin binds and ubiquitinates α-, β- and γ-tubulin. Integrity of the tubulin network is essential for the centrosome accumulation. It is speculated that the relocation of Parkin in the centrosome area facilitates the ubiquitination of its substrates when the cell encounters protein stress (Zhao et al. 2003; Ren et al. 2003). The mammalian RING-finger ubiquitin ligase Nrdp1 is able to degrade Parkin and thereby also regulates its activity (Zhong et al. 2005). Kalia et al. showed that the bcl-2-associated athanogene 5 (BAG-5) enhances dopamine neuron death in an in vivo model for Parkinson's disease through inhibition of the ubiquitin ligase activity of Parkin and the chaperone activity of Hsp70 (Kalia et al. 2004). Finally, the phosphorylation of five serines within Parkin probably also has a role in the regulation of the enzyme. In case of ER stress, phosphorylation of Parkin reduces, which increases its activity by decreasing its autoubiquitination (Yamamoto et al. 2005).

2.2.3.5
Malin

Lafora disease is a serious form of epilepsy affecting teenagers, usually causing death within 10 years of onset. It is characterized by an accumulation of starch-like polyglucosans in the ER of neuronal dendrites but not axons, called Lafora bodies (Minassian 2002). The epileptogenesis is linked to these ER-associated structures. Mutations in the *malin*, or NHLRC1, gene are linked to this disease (Chan et al. 2003). Malin has a classical RING-HC finger and six NHL repeat domains, which are protein–protein interaction domains (Slack and Ruvkun 1998). It localizes to the ER, and to a lesser extent to the nucleus, and it was present in all tissues tested (Chan et al. 2003). Two variant transcripts are being made, and close homologs are present in other vertebrates. Besides Malin, Laforin has also been linked to Lafora disease. Laforin is an ER-localized protein tyrosine phosphatase (Minassian et al. 1998, 2001). Judging from the ER associated accumulations that resemble features of Alzheimer's and Parkinson's diseases, it seems likely that Malin acts as a ubiquitin ligase in the degradation of ER proteins. This, however, has to be confirmed.

3
Substrate Specificity, Functional Redundancy, and Flexibility of Ubiquitin Ligases

Based on the data reviewed in the preceding paragraphs, one could conclude that E3 enzymes do not seem to be specific for either cytosolic or ER-associated substrates. Although it is tempting to speculate that ER-localized ubiquitin ligases such as Hrd1p, HRD1, and gp78 specifically induce the degradation of ER proteins, this may not be the case. This is illustrated by the degradation of cytosolic MATα2 under influence of the ER-integrated ubiquitin ligase Doa10p (Swanson et al. 2001). On the other hand, cytosolically localized E3 enzymes such as Rsp5p, CHIP, and Parkin are not restricted to the degradation of cytosolic proteins alone, but also serve a number of ER substrates. In this context, it is important to realize that the degradation of ER proteins involves two consecutive steps, i.e., dislocation and proteasomal degradation, each of which may be catalyzed by different E3 enzymes. Thus, there may be no formal difference between genuine cytosolic substrates and ER substrates that have been dislocated to the cytosol. This hypothesis may explain the observed involvement of cytosolically localized E3 ligases in the degradation of ER proteins.

The reviewed data also indicate that ubiquitin ligases have redundant functions, i.e., one protein can be degraded by a number of different E3 ligases. As described in earlier sections, CD3-δ can be degraded by HRD1 as well as by Gp78. Pael-R degradation is stimulated by both Parkin and CHIP, and possibly by HRD1. TCR-α can be targeted for degradation by Fbx2, Fbx6, and HRD1 ligases. CFTR degradation is influenced by CHIP and by the SCF^{Fbs1}ligase. The observed redundancy supposedly protects cells from the accumulation of misfolded proteins when one E3 enzyme cannot cope with the amount of substrate, or has otherwise become dysfunctional.

Besides having redundant functions, E3 ligases seem to be flexible with respect to their substrate populations. Ectopically expressed proteins that are foreign to the cell can still be served by one or more E3 ligases (Gnann et al. 2004), suggesting that substrate recognition may be flexible, responding to unique new requirements.

Flexibility is also illustrated by the fact that an E3 ligase can be forced to change its substrate population by binding another protein. β-TrCP, an SCF linked F-box protein, can be modified by HIV encoded Vpu to recognize CD4 and cause its ubiquitination, while CD4 is not normally a substrate of $SCF^{\beta TrCP}$. Also the HECT domain-containing ubiquitin ligase E6-AP does not normally degrade p53, but is forced to do so when it binds the oncoprotein E6 of human papilloma virus (Hengstermann et al. 2001).

4 Other Proteins with a Role in Endoplasmic Reticulum Degradation, Which May Interact with the Ubiquitination Machinery

The discovery of the degradation mechanism of ER proteins opened a new field of fundamental research. In addition, the finding that many common human diseases are linked to disturbed degradation of ER proteins has resulted in a substantial increase in publications on this subject in the past few years. After it was realized that the proteasome takes part in the degradation of ER proteins, an array of other proteins have recently been implicated in this degradation pathway. Below, a number of proteins are described that may cooperate with the ubiquitin enzymes discussed in this review to dislocate and degrade proteins from the ER.

In yeast, Hrd1p associates with Hrd3p, which assures Hrd1p's stability (Gardner et al. 2000). Hrd1p is autoregulated by self-ubiquitination and subsequent degradation. However, this is prevented by the presence of Hrd3p. Hrd3p somehow senses the presence of substrates at the lumenal side of the ER membrane and binds Hrd1p via the transmembrane region (Gardner et al. 2000). In humans, a homolog of Hrd3p has been identified, named SEL1L, which is highly expressed in the pancreas (Cattaneo et al. 2001). The *sel1* gene is conserved in many species, suggesting an important function (Biunno et al. 2002). Lowered expression of SEL1L in breast and pancreatic cancers is correlated with dramatic prognoses, suggesting that SEL1L inhibits tumor growth and aggressiveness (Orlandi et al. 2002; Cattaneo et al. 2003, 2004). Like HRD1, SEL1L is up-regulated as a result of ER stress, which may implicate that SEL1L indeed has a role in the degradation of ER proteins (Kaneko and Nomura 2003).

A factor that has been implicated in the degradation of all ER degradation substrates tested to date is the P97-Ufd1-Nlp4 complex (Ye et al. 2001; Jarosch et al. 2002; Rabinovich et al. 2002; Ye et al. 2003; Elkabetz et al. 2004). In yeast and mammalians, p97 is a cytosolic AAA-ATPase that binds to Ufd1 and Nlp4. In this configuration, p97 recognizes ubiquitinated as well as nonubiquitinated substrates at the ER membrane (Meyer et al. 2002; Ye et al. 2003). The ATPase activity of p97 is essential for the dislocation of ER proteins to the cytosol (Ye et al. 2003).

On their way to the cytosol, ER degradation substrates lose their N-linked glycans by the activity of a specialized peptide:N-glycanase located in the cytosol (Suzuki et al. 1998; Hirsch et al. 2003; Blom et al. 2004; Katiyar et al. 2004).

Derlin-1, the mammalian homolog of *S. cerevisiae* protein Der1p, also participates in the complex at the ER membrane that directs the dislocation of ER proteins (Ye et al. 2004; Lilley and Ploegh 2004). Derlin-1 binds to VIMP, a membrane-bound protein that recruits the p97-Ufd1-Nlp4 complex (Ye et al. 2004). Derlin-1 seems to be important only to a subset of ER degradation substrates, as is the case with its yeast homolog Der1p (Taxis et al. 2003; Lilley and Ploegh 2004).

HERP is another protein with a role in the degradation of ER proteins and neutralization of ER stress (Hori et al. 2004). This protein is anchored in the ER membrane and contains an N-terminal ubiquitin-like (UBL) domain.

HERP is significantly up-regulated upon ER stress due to UPR recognition sequences (both ERSE and ERSE II) in its promotor (Kokame et al. 2000; van Laar et al. 2000; Yamamoto et al. 2004; Hori et al. 2004). It also reacts to other forms of cellular stress such as amino acid deprivation, dsRNA expression as a result of viral infections, and heme deficiency (Ma and Hendershot 2004).

The recognition by the proteasome of a substrate that has been dislocated, deglycosylated, and ubiquitinated, is facilitated by RAD23, Ufd2, Rpn10, and/or Dsk2p in yeast. These proteins each contain ubiquitin-binding domains with which they bind ubiquitin-conjugated proteins and escort them from the ER membrane to the proteasome. (Wilkinson et al. 2001; Chen and Madura 2002; Elsasser et al. 2004; Verma et al. 2004; Richly et al. 2005).

It is likely that the factors described here form complexes that coordinate dislocation, deglycosylation, and ubiquitination of substrates, and target them for degradation by the proteasome. The stoichiometry and the precise mode of action of this multicomponent assembly remain to be established.

Acknowledgements We thank Martine Barel and Otto Ede Pool for their helpful discussions.

References

Aasland R, Gibson TJ, Stewart AF (1995) The PHD finger: implications for chromatin-mediated transcriptional regulation. Trends Biochem Sci 20:56–59

Alberti S, Demand J, Esser C, Emmerich N, Schild H, Hohfeld J (2002) Ubiquitylation of BAG-1 suggests a novel regulatory mechanism during the sorting of chaperone substrates to the proteasome. J Biol Chem 277:45920–45927

Allen JR, Nguyen LX, Sargent KE, Lipson KL, Hackett A, Urano F (2004) High ER stress in beta-cells stimulates intracellular degradation of misfolded insulin. Biochem Biophys Res Commun 324:166–170

Amano T, Yamasaki S, Yagishita N, Tsuchimochi K, Shin H, Kawahara K, Aratani S, Fujita H, Zhang L, Ikeda R, Fujii R, Miura N, Komiya S, Nishioka K, Maruyama I, Fukamizu A, Nakajima T (2003) Synoviolin/Hrd1, an E3 ubiquitin ligase, as a novel pathogenic factor for arthropathy. Genes Dev 17:2436–2449

Aravind L, Iyer LM, Koonin EV (2003) Scores of RINGS but no PHDs in ubiquitin signaling. Cell Cycle 2:123–126

Aravind L, Koonin EV (2000) The U box is a modified RING finger—a common domain in ubiquitination. Curr Biol 10:R132–R134

Ballinger CA, Connell P, Wu Y, Hu Z, Thompson LJ, Yin LY, Patterson C (1999) Identification of CHIP, a novel tetratricopeptide repeat-containing protein that interacts with heat shock proteins and negatively regulates chaperone functions. Mol Cell Biol 19:4535–4545

Bartee E, Mansouri M, Hovey Nerenberg BT, Gouveia K, Fruh K (2004) Downregulation of major histocompatibility complex class I by human ubiquitin ligases related to viral immune evasion proteins. J Virol 78:1109–1120

Bays NW, Gardner RG, Seelig LP, Joazeiro CA, Hampton RY (2001) Hrd1p/Der3p is a membrane-anchored ubiquitin ligase required for ER-associated degradation. Nat Cell Biol 3:24–29

Biederer T, Volkwein C, Sommer T (1996) Degradation of subunits of the Sec61p complex, an integral component of the ER membrane, by the ubiquitin-proteasome pathway. EMBO J 15:2069–2076

Biederer T, Volkwein C, Sommer T (1997) Role of Cue1p in ubiquitination and degradation at the ER surface. Science 278:1806–1809

Biunno I, Castiglioni B, Rogozin IB, DeBellis G, Malferrari G, Cattaneo M (2002) Cross-species conservation of SEL1L, a human pancreas-specific expressing gene. OMICS 6:187–198

Blom D, Hirsch C, Stern P, Tortorella D, Ploegh HL (2004) A glycosylated type I membrane protein becomes cytosolic when peptide: N-glycanase is compromised. EMBO J 23:650–658

Bordallo J, Plemper RK, Finger A, Wolf DH (1998) Der3p/Hrd1p is required for endoplasmic reticulum-associated degradation of misfolded lumenal and integral membrane proteins. Mol Biol Cell 9:209–222

Bordallo J, Wolf DH (1999) A RING-H2 finger motif is essential for the function of Der3/Hrd1 in endoplasmic reticulum associated protein degradation in the yeast Saccharomyces cerevisiae. FEBS Lett 448:244–248

Botero D, Gereben B, Goncalves C, De Jesus LA, Harney JW, Bianco AC (2002) Ubc6p and ubc7p are required for normal and substrate-induced endoplasmic reticulum-associated degradation of the human selenoprotein type 2 iodothyronine monodeiodinase. Mol Endocrinol 16:1999–2007

Bour S, Perrin C, Akari H, Strebel K (2001) The human immunodeficiency virus type 1 Vpu protein inhibits NF-kappa B activation by interfering with beta TrCP-mediated degradation of Ikappa B. J Biol Chem 276:15920–15928

Bruijn LI, Miller TM, Cleveland DW (2004) Unraveling the mechanisms involved in motor neuron degeneration in ALS. Annu Rev Neurosci 27:723–749

Busino L, Donzelli M, Chiesa M, Guardavaccaro D, Ganoth D, Dorrello NV, Hershko A, Pagano M, Draetta GF (2003) Degradation of Cdc25A by beta-TrCP during S phase and in response to DNA damage. Nature 426:87–91

Cattaneo M, Canton C, Albertini A, Biunno I (2004) Identification of a region within SEL1L protein required for tumour growth inhibition. Gene 326:149–156

Cattaneo M, Orlandini S, Beghelli S, Moore PS, Sorio C, Bonora A, Bassi C, Talamini G, Zamboni G, Orlandi R, Menard S, Bernardi LR, Biunno I, Scarpa A (2003) SEL1L expression in pancreatic adenocarcinoma parallels SMAD4 expression and delays tumor growth in vitro and in vivo. Oncogene 22:6359–6368

Cattaneo M, Sorio C, Malferrari G, Rogozin IB, Bernard L, Scarpa A, Zollo M, Biunno I (2001) Cloning and functional analysis of SEL1L promoter region, a pancreas-specific gene. DNA Cell Biol 20:1–9

Cesari R, Martin ES, Calin GA, Pentimalli F, Bichi R, McAdams H, Trapasso F, Drusco A, Shimizu M, Masciullo V, D'Andrilli G, Scambia G, Picchio MC, Alder H, Godwin AK, Croce CM (2003) Parkin, a gene implicated in autosomal recessive juvenile parkinsonism, is a candidate tumor suppressor gene on chromosome 6q25-q27. Proc Natl Acad Sci U S A 100:5956–5961

Chan EM, Young EJ, Ianzano L, Munteanu I, Zhao X, Christopoulos CC, Avanzini G, Elia M, Ackerley CA, Jovic NJ, Bohlega S, Andermann E, Rouleau GA, Delgado-Escueta AV, Minassian BA, Scherer SW (2003) Mutations in NHLRC1 cause progressive myoclonus epilepsy. Nat Genet 35:125–127

Chang A, Cheang S, Espanel X, Sudol M (2000) Rsp5 WW domains interact directly with the carboxyl-terminal domain of RNA polymerase II. J Biol Chem 275:20562–20571

Chen L, Madura K (2002) Rad23 promotes the targeting of proteolytic substrates to the proteasome. Mol Cell Biol 22:4902–4913

Chung KK, Zhang Y, Lim KL, Tanaka Y, Huang H, Gao J, Ross CA, Dawson VL, Dawson TM (2001) Parkin ubiquitinates the alpha-synuclein-interacting protein, synphilin-1: implications for Lewy-body formation in Parkinson disease. Nat Med 7:1144–1150

Connell P, Ballinger CA, Jiang J, Wu Y, Thompson LJ, Hohfeld J, Patterson C (2001) The co-chaperone CHIP regulates protein triage decisions mediated by heat-shock proteins. Nat Cell Biol 3:93–96

Corti O, Hampe C, Koutnikova H, Darios F, Jacquier S, Prigent A, Robinson JC, Pradier L, Ruberg M, Mirande M, Hirsch E, Rooney T, Fournier A, Brice A (2003) The p38 subunit of the aminoacyl-tRNA synthetase complex is a Parkin substrate: linking protein biosynthesis and neurodegeneration. Hum Mol Genet 12:1427–1437

Coscoy L, Ganem D (2000) Kaposi's sarcoma-associated herpesvirus encodes two proteins that block cell surface display of MHC class I chains by enhancing their endocytosis. Proc Natl Acad Sci U S A 97:8051–8056

Coscoy L, Ganem D (2001) A viral protein that selectively downregulates ICAM-1 and B7-2 and modulates T cell costimulation. J Clin Invest 107:1599–1606

Coscoy L, Sanchez DJ, Ganem D (2001) A novel class of herpesvirus-encoded membrane-bound E3 ubiquitin ligases regulates endocytosis of proteins involved in immune recognition. J Cell Biol 155:1265–1273

De Virgilio M, Weninger H, Ivessa NE (1998) Ubiquitination is required for the retro-translocation of a short-lived luminal endoplasmic reticulum glycoprotein to the cytosol for degradation by the proteasome. J Biol Chem 273:9734–9743

Deak PM, Wolf DH (2001) Membrane topology and function of Der3/Hrd1p as a ubiquitin-protein ligase (E3) involved in endoplasmic reticulum degradation. J Biol Chem 276:10663–10669

Demand J, Alberti S, Patterson C, Hohfeld J (2001) Cooperation of a ubiquitin domain protein and an E3 ubiquitin ligase during chaperone/proteasome coupling. Curr Biol 11:1569–1577

Denison SR, Wang F, Becker NA, Schule B, Kock N, Phillips LA, Klein C, Smith DI (2003) Alterations in the common fragile site gene Parkin in ovarian and other cancers. Oncogene 22:8370–8378

Deshaies RJ (1999) SCF and Cullin/Ring H2-based ubiquitin ligases. Annu Rev Cell Dev Biol 15:435–467

Elkabetz Y, Shapira I, Rabinovich E, Bar-Nun S (2004) Distinct steps in dislocation of luminal endoplasmic reticulum-associated degradation substrates: roles of endoplasmic reticulum-bound p97/Cdc48p and proteasome. J Biol Chem 279:3980–3989

Elsasser S, Chandler-Militello D, Muller B, Hanna J, Finley D (2004) Rad23 and Rpn10 serve as alternative ubiquitin receptors for the proteasome. J Biol Chem 279:26817–26822

Fang S, Ferrone M, Yang C, Jensen JP, Tiwari S, Weissman AM (2001) The tumor autocrine motility factor receptor, gp78, is a ubiquitin protein ligase implicated in degradation from the endoplasmic reticulum. Proc Natl Acad Sci U S A 98:14422–14427

Fang S, Weissman AM (2004) A field guide to ubiquitylation. Cell Mol Life Sci 61:1546–1561

Finney N, Walther F, Mantel PY, Stauffer D, Rovelli G, Dev KK (2003) The cellular protein level of Parkin is regulated by its ubiquitin-like domain. J Biol Chem 278:16054–16058

Friedlander R, Jarosch E, Urban J, Volkwein C, Sommer T (2000) A regulatory link between ER-associated protein degradation and the unfolded-protein response. Nat Cell Biol 2:379–384

Galan JM, Moreau V, Andre B, Volland C, Haguenauer-Tsapis R (1996) Ubiquitination mediated by the Npi1p/Rsp5p ubiquitin-protein ligase is required for endocytosis of the yeast uracil permease. J Biol Chem 271:10946–10952

Gardner RG, Shearer AG, Hampton RY (2001) In vivo action of the HRD ubiquitin ligase complex: mechanisms of endoplasmic reticulum quality control and sterol regulation. Mol Cell Biol 21:4276–4291

Gardner RG, Swarbrick GM, Bays NW, Cronin SR, Wilhovsky S, Seelig L, Kim C, Hampton RY (2000) Endoplasmic reticulum degradation requires lumen to cytosol signaling. Transmembrane control of Hrd1p by Hrd3p. J Cell Biol 151:69–82

Glass AS, Huynh DP, Franck T, Woitalla D, Muller T, Pulst SM, Berg D, Kruger R, Riess O (2004) Screening for mutations in synaptotagmin XI in Parkinson's disease. J Neural Transm Suppl 68:21–28

Gnann A, Riordan JR, Wolf DH (2004) Cystic fibrosis transmembrane conductance regulator degradation depends on the lectins Htm1p/EDEM and the Cdc48 protein complex in yeast. Mol Biol Cell 15:4125–4135

Goldberg MS, Fleming SM, Palacino JJ, Cepeda C, Lam HA, Bhatnagar A, Meloni EG, Wu N, Ackerson LC, Klapstein GJ, Gajendiran M, Roth BL, Chesselet MF, Maidment NT, Levine MS, Shen J (2003) Parkin-deficient mice exhibit nigrostriatal deficits but not loss of dopaminergic neurons. J Biol Chem 278:43628–43635

Guardavaccaro D, Kudo Y, Boulaire J, Barchi M, Busino L, Donzelli M, Margottin-Goguet F, Jackson PK, Yamasaki L, Pagano M (2003) Control of meiotic and mitotic progression by the F box protein beta-Trcp1 in vivo. Dev Cell 4:799–812

Hampton RY (2002) Proteolysis and sterol regulation. Annu Rev Cell Dev Biol 18:345–378

Hampton RY, Gardner RG, Rine J (1996) Role of 26S proteasome and HRD genes in the degradation of 3-hydroxy-3-methylglutaryl-CoA reductase, an integral endoplasmic reticulum membrane protein. Mol Biol Cell 7:2029–2044

Harris A, Argent BE (1993) The cystic fibrosis gene and its product CFTR. Semin Cell Biol 4:37–44

Hart M, Concordet JP, Lassot I, Albert I, del los Santos R, Durand H, Perret C, Rubinfeld B, Margottin F, Benarous R, Polakis P (1999) The F-box protein beta-TrCP associates with phosphorylated beta-catenin and regulates its activity in the cell. Curr Biol 9:207–210

Hassink GC, Kikkert M, van Voorden S, Lee SJ, Spaapen R, van Laar T, Coleman CS, Bartee E, Fruh K, Chau V, Wiertz EJ (2005) TEB4 is a C4HC3 RING finger-containing ubiquitin ligase of the endoplasmic reticulum. Biochem J (in press)

Hatakeyama S, Matsumoto M, Kamura T, Murayama M, Chui DH, Planel E, Takahashi R, Nakayama KI, Takashima A (2004a) U-box protein carboxyl terminus of Hsc70-interacting protein (CHIP) mediates poly-ubiquitylation preferentially on four-repeat Tau and is involved in neurodegeneration of tauopathy. J Neurochem 91:299–307

Hatakeyama S, Matsumoto M, Yada M, Nakayama KI (2004b) Interaction of U-box-type ubiquitin-protein ligases (E3s) with molecular chaperones. Genes Cells 9:533–548

Haynes CM, Caldwell S, Cooper AA (2002) An HRD/DER-independent ER quality control mechanism involves Rsp5p-dependent ubiquitination and ER-Golgi transport. J Cell Biol 158:91–101

Hein C, Springael JY, Volland C, Haguenauer-Tsapis R, Andre B (1995) NPl1, an essential yeast gene involved in induced degradation of Gap1 and Fur4 permeases, encodes the Rsp5 ubiquitin-protein ligase. Mol Microbiol 18:77–87

Hengstermann A, Linares LK, Ciechanover A, Whitaker NJ, Scheffner M (2001) Complete switch from Mdm2 to human papillomavirus E6-mediated degradation of p53 in cervical cancer cells. Proc Natl Acad Sci U S A 98:1218–1223

Hershko A, Ciechanover A (1998) The ubiquitin system. Annu Rev Biochem 67:425–479

Hewitt EW, Duncan L, Mufti D, Baker J, Stevenson PG, Lehner PJ (2002) Ubiquitylation of MHC class I by the K3 viral protein signals internalization and TSG101-dependent degradation. EMBO J 21:2418–2429

Hicke L (1999) Gettin' down with ubiquitin: turning off cell-surface receptors, transporters and channels. Trends Cell Biol 9:107–112

Hicke L, Dunn R (2003) Regulation of membrane protein transport by ubiquitin and ubiquitin-binding proteins. Annu Rev Cell Dev Biol 19:141–172

Hill K, Cooper AA (2000) Degradation of unassembled Vph1p reveals novel aspects of the yeast ER quality control system. EMBO J 19:550–561

Hiller MM, Finger A, Schweiger M, Wolf DH (1996) ER degradation of a misfolded luminal protein by the cytosolic ubiquitin-proteasome pathway. Science 273:1725–1728

Hirono Y, Fushida S, Yonemura Y, Yamamoto H, Watanabe H, Raz A (1996) Expression of autocrine motility factor receptor correlates with disease progression in human gastric cancer. Br J Cancer 74:2003–2007

Hirsch C, Blom D, Ploegh HL (2003) A role for N-glycanase in the cytosolic turnover of glycoproteins. EMBO J 22:1036–1046

Hirsch C, Jarosch E, Sommer T, Wolf DH (2004) Endoplasmic reticulum-associated protein degradation—one model fits all? Biochim Biophys Acta 1695:215–223

Hitchcock AL, Auld K, Gygi SP, Silver PA (2003) A subset of membrane-associated proteins is ubiquitinated in response to mutations in the endoplasmic reticulum degradation machinery. Proc Natl Acad Sci U S A 100:12735–12740

Hoppe T, Matuschewski K, Rape M, Schlenker S, Ulrich HD, Jentsch S (2000) Activation of a membrane-bound transcription factor by regulated ubiquitin/proteasome-dependent processing. Cell 102:577–586

Hori O, Ichinoda F, Yamaguchi A, Tamatani T, Taniguchi M, Koyama Y, Katayama T, Tohyama M, Stern DM, Ozawa K, Kitao Y, Ogawa S (2004) Role of Herp in the endoplasmic reticulum stress response. Genes Cells 9:457–469

Hoshikawa C, Shichiri M, Nakamori S, Takagi H (2003) A nonconserved Ala401 in the yeast Rsp5 ubiquitin ligase is involved in degradation of Gap1 permease and stress-induced abnormal proteins. Proc Natl Acad Sci U S A 100:11505–11510

Huang Z, Nie L, Xu M, Sun XH (2004) Notch-induced E2A degradation requires CHIP and Hsc70 as novel facilitators of ubiquitination. Mol Cell Biol 24:8951–8962

Huibregtse JM, Scheffner M, Beaudenon S, Howley PM (1995) A family of proteins structurally and functionally related to the E6-AP ubiquitin-protein ligase. Proc Natl Acad Sci U S A 92:2563–2567

Huynh DP, Scoles DR, Nguyen D, Pulst SM (2003) The autosomal recessive juvenile Parkinson disease gene product, Parkin, interacts with and ubiquitinates synaptotagmin XI. Hum Mol Genet 12:2587–2597

Imai Y, Soda M, Hatakeyama S, Akagi T, Hashikawa T, Nakayama KI, Takahashi R (2002) CHIP is associated with Parkin, a gene responsible for familial Parkinson's disease, and enhances its ubiquitin ligase activity. Mol Cell 10:55–67

Imai Y, Soda M, Inoue H, Hattori N, Mizuno Y, Takahashi R (2001) An unfolded putative transmembrane polypeptide, which can lead to endoplasmic reticulum stress, is a substrate of Parkin. Cell 105:891–902

Imai Y, Soda M, Takahashi R (2000) Parkin suppresses unfolded protein stress-induced cell death through its E3 ubiquitin-protein ligase activity. J Biol Chem 275:35661–35664

Ishido S, Choi JK, Lee BS, Wang C, DeMaria M, Johnson RP, Cohen GB, Jung JU (2000a) Inhibition of natural killer cell-mediated cytotoxicity by Kaposi's sarcoma-associated herpesvirus K5 protein. Immunity 13:365–374

Ishido S, Wang C, Lee BS, Cohen GB, Jung JU (2000b) Downregulation of major histocompatibility complex class I molecules by Kaposi's sarcoma-associated herpesvirus K3 and K5 proteins. J Virol 74:5300–5309

Ivan M, Kondo K, Yang H, Kim W, Valiando J, Ohh M, Salic A, Asara JM, Lane WS, Kaelin WG Jr (2001) HIFalpha targeted for VHL-mediated destruction by proline hydroxylation: implications for O2 sensing. Science 292:464–468

Jaakkola P, Mole DR, Tian YM, Wilson MI, Gielbert J, Gaskell SJ, Kriegsheim A, Hebestreit HF, Mukherji M, Schofield CJ, Maxwell PH, Pugh CW, Ratcliffe PJ (2001) Targeting of HIF-alpha to the von Hippel-Lindau ubiquitylation complex by O2-regulated prolyl hydroxylation. Science 292:468–472

Jarosch E, Taxis C, Volkwein C, Bordallo J, Finley D, Wolf DH, Sommer T (2002) Protein dislocation from the ER requires polyubiquitination and the AAA-ATPase Cdc48. Nat Cell Biol 4:134–139

Jiang J, Ballinger CA, Wu Y, Dai Q, Cyr DM, Hohfeld J, Patterson C (2001) CHIP is a U-box-dependent E3 ubiquitin ligase: identification of Hsc70 as a target for ubiquitylation. J Biol Chem 276:42938–42944

Kalia SK, Lee S, Smith PD, Liu L, Crocker SJ, Thorarinsdottir TE, Glover JR, Fon EA, Park DS, Lozano AM (2004) BAG5 inhibits parkin and enhances dopaminergic neuron degeneration. Neuron 44:931–945

Kaneko M, Ishiguro M, Niinuma Y, Uesugi M, Nomura Y (2002) Human HRD1 protects against ER stress-induced apoptosis through ER-associated degradation. FEBS Lett 532:147–152

Kaneko M, Nomura Y (2003) ER signaling in unfolded protein response. Life Sci 74:199–205

Kaneko M, Nomura Y (2004) Protective effects of HRD1 and 4-phenylbutyric acid against neuronal cell death. Nippon Yakurigaku Zasshi 124:391–398

Katiyar S, Li G, Lennarz WJ (2004) A complex between peptide:N-glycanase and two proteasome-linked proteins suggests a mechanism for the degradation of misfolded glycoproteins. Proc Natl Acad Sci U S A 101:13774–13779

Katzmann DJ, Sarkar S, Chu T, Audhya A, Emr SD (2004) Multivesicular body sorting: ubiquitin ligase Rsp5 is required for the modification and sorting of carboxypeptidase S. Mol Biol Cell 15:468–480

Kikkert M, Doolman R, Dai M, Avner R, Hassink G, van Voorden S, Thanedar S, Roitelman J, Chau V, Wiertz E (2004) Human HRD1 is an E3 ubiquitin ligase involved in degradation of proteins from the endoplasmic reticulum. J Biol Chem 279:3525–3534

Kikkert M, Hassink G, Barel M, Hirsch C, van der Wal FJ, Wiertz E (2001) Ubiquitination is essential for human cytomegalovirus US11-mediated dislocation of MHC class I molecules from the endoplasmic reticulum to the cytosol. Biochem J 358:369–377

Kim BW, Zavacki AM, Curcio-Morelli C, Dentice M, Harney JW, Larsen PR, Bianco AC (2003) Endoplasmic reticulum-associated degradation of the human type 2 iodothyronine deiodinase (D2) is mediated via an association between mammalian UBC7 and the carboxyl region of D2. Mol Endocrinol 17:2603–2612

Kipreos ET, Pagano M (2000) The F-box protein family. Genome Biol 1:REVIEWS3002

Kiser GL, Gentzsch M, Kloser AK, Balzi E, Wolf DH, Goffeau A, Riordan JR (2001) Expression and degradation of the cystic fibrosis transmembrane conductance regulator in Saccharomyces cerevisiae. Arch Biochem Biophys 390:195–205

Kitada T, Asakawa S, Hattori N, Matsumine H, Yamamura Y, Minoshima S, Yokochi M, Mizuno Y, Shimizu N (1998) Mutations in the Parkin gene cause autosomal recessive juvenile parkinsonism. Nature 392:605–608

Kitada T, Asakawa S, Minoshima S, Mizuno Y, Shimizu N (2000) Molecular cloning, gene expression, and identification of a splicing variant of the mouse Parkin gene. Mamm Genome 11:417–421

Kokame K, Agarwala KL, Kato H, Miyata T (2000) Herp, a new ubiquitin-like membrane protein induced by endoplasmic reticulum stress. J Biol Chem 275:32846–32853

Kosarev P, Mayer KF, Hardtke CS (2002) Evaluation and classification of RING-finger domains encoded by the Arabidopsis genome. Genome Biol 3:RESEARCH0016

Lenk U, Yu H, Walter J, Gelman MS, Hartmann E, Kopito RR, Sommer T (2002) A role for mammalian Ubc6 homologues in ER-associated protein degradation. J Cell Sci 115:3007–3014

Liang JS, Kim T, Fang S, Yamaguchi J, Weissman AM, Fisher EA, Ginsberg HN (2003) Overexpression of the tumor autocrine motility factor receptor Gp78, a ubiquitin protein ligase, results in increased ubiquitinylation and decreased secretion of apolipoprotein B100 in HepG2 cells. J Biol Chem 278:23984–23988

Lilley BN, Ploegh HL (2004) A membrane protein required for dislocation of misfolded proteins from the ER. Nature 429:834–840

Liu C, Kato Y, Zhang Z, Do VM, Yankner BA, He X (1999) beta-Trcp couples beta-catenin phosphorylation-degradation and regulates Xenopus axis formation. Proc Natl Acad Sci U S A 96:6273–6278

Lorenzetti D, Antalffy B, Vogel H, Noveroske J, Armstrong D, Justice M (2004) The neurological mutant quaking(viable) is Parkin deficient. Mamm Genome 15:210–217

Lorenzo ME, Jung JU, Ploegh HL (2002) Kaposi's sarcoma-associated herpesvirus K3 utilizes the ubiquitin-proteasome system in routing class major histocompatibility complexes to late endocytic compartments. J Virol 76:5522–5531

Lorick KL, Jensen JP, Fang S, Ong AM, Hatakeyama S, Weissman AM (1999) RING fingers mediate ubiquitin-conjugating enzyme (E2)-dependent ubiquitination. Proc Natl Acad Sci U S A 96:11364–11369

Luders J, Demand J, Hohfeld J (2000) The ubiquitin-related BAG-1 provides a link between the molecular chaperones Hsc70/Hsp70 and the proteasome. J Biol Chem 275:4613–4617

Lyngso C, Bouteiller G, Damgaard CK, Ryom D, Sanchez-Munoz S, Norby PL, Bonven BJ, Jorgensen P (2000) Interaction between the transcription factor SPBP and the positive cofactor RNF4. An interplay between protein binding zinc fingers. J Biol Chem 275:26144–26149

Ma Y, Hendershot LM (2004) Herp is dually regulated by both the endoplasmic reticulum stress-specific branch of the unfolded protein response and a branch that is shared with other cellular stress pathways. J Biol Chem 279:13792–13799

Margottin F, Bour SP, Durand H, Selig L, Benichou S, Richard V, Thomas D, Strebel K, Benarous R (1998) A novel human WD protein, h-beta TrCp, that interacts with HIV-1 Vpu connects CD4 to the ER degradation pathway through an F-box motif. Mol Cell 1:565–574

Margottin-Goguet F, Hsu JY, Loktev A, Hsieh HM, Reimann JD, Jackson PK (2003) Prophase destruction of Emi1 by the SCF(betaTrCP/Slimb) ubiquitin ligase activates the anaphase promoting complex to allow progression beyond prometaphase. Dev Cell 4:813–826

Marmor MD, Yarden Y (2004) Role of protein ubiquitylation in regulating endocytosis of receptor tyrosine kinases. Oncogene 23:2057–2070

Maruyama K, Watanabe H, Shiozaki H, Takayama T, Gofuku J, Yano H, Inoue M, Tamura S, Raz A, Monden M (1995) Expression of autocrine motility factor receptor in human esophageal squamous cell carcinoma. Int J Cancer 64:316–321

McDonough H, Patterson C (2003) CHIP: a link between the chaperone and proteasome systems. Cell Stress Chaperones 8:303–308

McNaught KS, Olanow CW (2003) Proteolytic stress: a unifying concept for the etiopathogenesis of Parkinson's disease. Ann Neurol 53 [Suppl 3]:S73–S84

Meacham GC, Patterson C, Zhang W, Younger JM, Cyr DM (2001) The Hsc70 co-chaperone CHIP targets immature CFTR for proteasomal degradation. Nat Cell Biol 3:100–105

Meyer HH, Wang Y, Warren G (2002) Direct binding of ubiquitin conjugates by the mammalian p97 adaptor complexes, p47 and Ufd1-Npl4. EMBO J 21:5645–5652

Minassian BA (2002) Progressive myoclonus epilepsy with polyglucosan bodies: Lafora disease. Adv Neurol 89:199–210

Minassian BA, Andrade DM, Ianzano L, Young EJ, Chan E, Ackerley CA, Scherer SW (2001) Laforin is a cell membrane and endoplasmic reticulum-associated protein tyrosine phosphatase. Ann Neurol 49:271–275

Minassian BA, Lee JR, Herbrick JA, Huizenga J, Soder S, Mungall AJ, Dunham I, Gardner R, Fong CY, Carpenter S, Jardim L, Satishchandra P, Andermann E, Snead OC, III, Lopes-Cendes I, Tsui LC, Delgado-Escueta AV, Rouleau GA, Scherer SW (1998) Mutations in a gene encoding a novel protein tyrosine phosphatase cause progressive myoclonus epilepsy. Nat Genet 20:171–174

Mizushima T, Hirao T, Yoshida Y, Lee SJ, Chiba T, Iwai K, Yamaguchi Y, Kato K, Tsukihara T, Tanaka K (2004) Structural basis of sugar-recognizing ubiquitin ligase. Nat Struct Mol Biol 11:365–370

Murakami T, Shoji M, Imai Y, Inoue H, Kawarabayashi T, Matsubara E, Harigaya Y, Sasaki A, Takahashi R, Abe K (2004) Pael-R is accumulated in Lewy bodies of Parkinson's disease. Ann Neurol 55:439–442

Nabi IR, Watanabe H, Raz A (1990) Identification of B16-F1 melanoma autocrine motility-like factor receptor. Cancer Res. 50:409–414

Nabi IR, Watanabe H, Raz A (1992) Autocrine motility factor and its receptor: role in cell locomotion and metastasis. Cancer Metastasis Rev 11:5–20

Nadav E, Shmueli A, Barr H, Gonen H, Ciechanover A, Reiss Y (2003) A novel mammalian endoplasmic reticulum ubiquitin ligase homologous to the yeast Hrd1. Biochem Biophys Res Commun 303:91–97

Nakamori S, Watanabe H, Kameyama M, Imaoka S, Furukawa H, Ishikawa O, Sasaki Y, Kabuto T, Raz A (1994) Expression of autocrine motility factor receptor in colorectal cancer as a predictor for disease recurrence. Cancer 74:1855–1862

Nakatani Y, Kaneto H, Kawamori D, Yoshiuchi K, Hatazaki M, Matsuoka TA, Ozawa K, Ogawa S, Hori M, Yamasaki Y, Matsuhisa M (2005) Involvement of endoplasmic reticulum stress in insulin resistance and diabetes. J Biol Chem 280:847–851

Onishi Y, Tsukada K, Yokota J, Raz A (2003) Overexpression of autocrine motility factor receptor (AMFR) in NIH3T3 fibroblasts induces cell transformation. Clin Exp Metastasis 20:51–58

Orlandi R, Cattaneo M, Troglio F, Casalini P, Ronchini C, Menard S, Biunno I (2002) SEL1L expression decreases breast tumor cell aggressiveness in vivo and in vitro. Cancer Res 62:567–574

Otto T, Bex A, Schmidt U, Raz A, Rubben H (1997) Improved prognosis assessment for patients with bladder carcinoma. Am J Pathol 150:1919–1923

Ozcan U, Cao Q, Yilmaz E, Lee AH, Iwakoshi NN, Ozdelen E, Tuncman G, Gorgun C, Glimcher LH, Hotamisligil GS (2004) Endoplasmic reticulum stress links obesity, insulin action, and type 2 diabetes. Science 306:457–461

Palacino JJ, Sagi D, Goldberg MS, Krauss S, Motz C, Wacker M, Klose J, Shen J (2004) Mitochondrial dysfunction and oxidative damage in Parkin-deficient mice. J Biol Chem 279:18614–18622

Peters JM (2003) Emi1 proteolysis: how SCF(beta-Trcp1) helps to activate the anaphase-promoting complex. Mol Cell 11:1420–1421

Petroski MD, Deshaies RJ (2005) Function and regulation of cullin-RING ubiquitin ligases. Nat Rev Mol Cell Biol 6:9–20

Plemper RK, Bohmler S, Bordallo J, Sommer T, Wolf DH (1997) Mutant analysis links the translocon and BiP to retrograde protein transport for ER degradation. Nature 388:891–895

Ponting CP (2000) Proteins of the endoplasmic-reticulum-associated degradation pathway: domain detection and function prediction. Biochem J 351:527–535

Qi X, Okuma Y, Hosoi T, Kaneko M, Nomura Y (2004) Induction of murine HRD1 in experimental cerebral ischemia. Brain Res Mol Brain Res 130:30–38

Rabinovich E, Kerem A, Frohlich KU, Diamant N, Bar-Nun S (2002) AAA-ATPase p97/Cdc48p, a cytosolic chaperone required for endoplasmic reticulum-associated protein degradation. Mol Cell Biol 22:626–634

Ren Y, Zhao J, Feng J (2003) Parkin binds to alpha/beta tubulin and increases their ubiquitination and degradation. J Neurosci 23:3316–3324

Richly H, Rape M, Braun S, Rumpf S, Hoege C, Jentsch S (2005) A series of ubiquitin binding factors connects CDC48/p97 to substrate multiubiquitylation and proteasomal targeting. Cell 120:73–84

Rodriguez MS, Gwizdek C, Haguenauer-Tsapis R, Dargemont C (2003) The HECT ubiquitin ligase Rsp5p is required for proper nuclear export of mRNA in Saccharomyces cerevisiae. Traffic. 4:566–575

Saha V, Chaplin T, Gregorini A, Ayton P, Young BD (1995) The leukemia-associated-protein (LAP) domain, a cysteine-rich motif, is present in a wide range of proteins, including MLL, AF10, and MLLT6 proteins. Proc Natl Acad Sci U S A 92:9737–9741

Sakata E, Yamaguchi Y, Kurimoto E, Kikuchi J, Yokoyama S, Yamada S, Kawahara H, Yokosawa H, Hattori N, Mizuno Y, Tanaka K, Kato K (2003) Parkin binds the Rpn10 subunit of 26S proteasomes through its ubiquitin-like domain. EMBO Rep 4:301–306

Schindler U, Beckmann H, Cashmore AR (1993) HAT3.1, a novel Arabidopsis homeodomain protein containing a conserved cysteine-rich region. Plant J 4:137–150

Shamu CE, Flierman D, Ploegh HL, Rapoport TA, Chau V (2001) Polyubiquitination is required for US11-dependent movement of MHC class I heavy chain from endoplasmic reticulum into cytosol. Mol Biol Cell 12:2546–2555

Shamu CE, Story CM, Rapoport TA, Ploegh HL (1999) The pathway of US11-dependent degradation of MHC class I heavy chains involves a ubiquitin-conjugated intermediate. J Cell Biol 147:45–58

Shcherbik N, Haines DS (2004) Ub on the move. J Cell Biochem 93:11–19

Shih SC, Prag G, Francis SA, Sutanto MA, Hurley JH, Hicke L (2003) A ubiquitin-binding motif required for intramolecular monoubiquitylation, the CUE domain. EMBO J 22:1273–1281

Shimizu K, Tani M, Watanabe H, Nagamachi Y, Niinaka Y, Shiroishi T, Ohwada S, Raz A, Yokota J (1999) The autocrine motility factor receptor gene encodes a novel type of seven transmembrane protein. FEBS Lett 456:295–300

Shimura H, Schlossmacher MG, Hattori N, Frosch MP, Trockenbacher A, Schneider R, Mizuno Y, Kosik KS, Selkoe DJ (2001) Ubiquitination of a new form of alpha-synuclein by Parkin from human brain: implications for Parkinson's disease. Science 293:263–269

Sigismund S, Polo S, Di Fiore PP (2004) Signaling through monoubiquitination. Curr Top Microbiol Immunol 286:149–185

Slack FJ, Ruvkun G (1998) A novel repeat domain that is often associated with RING finger and B-box motifs. Trends Biochem Sci 23:474–475

Sommer T, Jentsch S (1993) A protein translocation defect linked to ubiquitin conjugation at the endoplasmic reticulum. Nature 365:176–179

Sommer T, Wolf DH (1997) Endoplasmic reticulum degradation: reverse protein flow of no return. FASEB J 11:1227–1233

Stahl PD, Barbieri MA (2002) Multivesicular bodies and multivesicular endosomes: the "ins and outs" of endosomal traffic. Sci STKE 2002:E32

Staropoli JF, McDermott C, Martinat C, Schulman B, Demireva E, Abeliovich A (2003) Parkin is a component of an SCF-like ubiquitin ligase complex and protects postmitotic neurons from kainate excitotoxicity. Neuron 37:735–749

Strous GJ, van Kerkhof P (2002) The ubiquitin-proteasome pathway and the regulation of growth hormone receptor availability. Mol Cell Endocrinol 197:143–151

Suzuki T, Park H, Kitajima K, Lennarz WJ (1998) Peptides glycosylated in the endoplasmic reticulum of yeast are subsequently deglycosylated by a soluble peptide: N-glycanase activity. J Biol Chem 273:21526–21530

Swanson R, Locher M, Hochstrasser M (2001) A conserved ubiquitin ligase of the nuclear envelope/endoplasmic reticulum that functions in both ER-associated and Matalpha2 repressor degradation. Genes Dev 15:2660–2674

Takahashi R, Imai Y (2003) Pael receptor, endoplasmic reticulum stress, and Parkinson's disease. J Neurol 250 [Suppl 3]:III25–III29

Takahashi R, Imai Y, Hattori N, Mizuno Y (2003) Parkin and endoplasmic reticulum stress. Ann N Y Acad Sci 991:101–106

Taxis C, Hitt R, Park SH, Deak PM, Kostova Z, Wolf DH (2003) Use of modular substrates demonstrates mechanistic diversity and reveals differences in chaperone requirement of ERAD. J Biol Chem 278:35903–35913

Tiwari S, Weissman AM (2001) Endoplasmic reticulum (ER)-associated degradation of T cell receptor subunits. Involvement of ER-associated ubiquitin-conjugating enzymes (E2s). J Biol Chem 276:16193–16200

Tsai YC, Fishman PS, Thakor NV, Oyler GA (2003) Parkin facilitates the elimination of expanded polyglutamine proteins and leads to preservation of proteasome function. J Biol Chem 278:22044–22055

Urushitani M, Kurisu J, Tateno M, Hatakeyama S, Nakayama K, Kato S, Takahashi R (2004) CHIP promotes proteasomal degradation of familial ALS-linked mutant SOD1 by ubiquitinating Hsp/Hsc70. J Neurochem 90:231–244

Van Laar T, Schouten T, Hoogervorst E, van Eck M, van der Eb AJ, Terleth C (2000) The novel MMS-inducible gene Mif1/KIAA0025 is a target of the unfolded protein response pathway. FEBS Lett 469:123–131

Vashist S, Ng DT (2004) Misfolded proteins are sorted by a sequential checkpoint mechanism of ER quality control. J Cell Biol 165:41–52

Verma R, Oania R, Graumann J, Deshaies RJ (2004) Multiubiquitin chain receptors define a layer of substrate selectivity in the ubiquitin-proteasome system. Cell 118:99–110

Walter J, Urban J, Volkwein C, Sommer T (2001) Sec61p-independent degradation of the tail-anchored ER membrane protein Ubc6p. EMBO J 20:3124–3131

Wang G, Yang J, Huibregtse JM (1999) Functional domains of the Rsp5 ubiquitin-protein ligase. Mol Cell Biol 19:342–352

Wang M, Suzuki T, Kitada T, Asakawa S, Minoshima S, Shimizu N, Tanaka K, Mizuno Y, Hattori N (2001) Developmental changes in the expression of Parkin and UbcR7, a Parkin-interacting and ubiquitin-conjugating enzyme, in rat brain. J Neurochem 77:1561–1568

Wang Q, Chang A (2003) Substrate recognition in ER-associated degradation mediated by Eps1, a member of the protein disulfide isomerase family. EMBO J 22:3792–3802

Ward CL, Omura S, Kopito RR (1995) Degradation of CFTR by the ubiquitin-proteasome pathway. Cell 83:121–127

Watanabe H, Nabi IR, Raz A (1991) The relationship between motility factor receptor internalization and the lung colonization capacity of murine melanoma cells. Cancer Res 51:2699–2705

Webster JM, Tiwari S, Weissman AM, Wojcikiewicz RJ (2003) Inositol 1,4,5-trisphosphate receptor ubiquitination is mediated by mammalian Ubc7, a component of the endoplasmic reticulum-associated degradation pathway, and is inhibited by chelation of intracellular Zn2+. J Biol Chem 278:38238–38246

Weissman AM (2001) Themes and variations on ubiquitylation. Nat Rev Mol Cell Biol 2:169–178

West AB, Kapatos G, O'Farrell C, Gonzalez-de-Chavez F, Chiu K, Farrer MJ, Maidment NT (2004) N-myc regulates Parkin expression. J Biol Chem 279:28896–28902

Wiertz EJ, Jones TR, Sun L, Bogyo M, Geuze HJ, Ploegh HL (1996a) The human cytomegalovirus US11 gene product dislocates MHC class I heavy chains from the endoplasmic reticulum to the cytosol. Cell 84:769–779

Wiertz EJ, Tortorella D, Bogyo M, Yu J, Mothes W, Jones TR, Rapoport TA, Ploegh HL (1996b) Sec61-mediated transfer of a membrane protein from the endoplasmic reticulum to the proteasome for destruction. Nature 384:432–438

Wilhovsky S, Gardner R, Hampton R (2000) HRD gene dependence of endoplasmic reticulum-associated degradation. Mol Biol Cell 11:1697–1708

Wilkinson CR, Seeger M, Hartmann-Petersen R, Stone M, Wallace M, Semple C, Gordon C (2001) Proteins containing the UBA domain are able to bind to multi-ubiquitin chains. Nat Cell Biol 3:939–943

Winston JT, Koepp DM, Zhu C, Elledge SJ, Harper JW (1999) A family of mammalian F-box proteins. Curr Biol 9:1180–1182

Yamamoto A, Friedlein A, Imai Y, Takahashi R, Kahle PJ, Haass C (2005) Parkin phosphorylation and modulation of its E3 ubiquitin ligase activity. J Biol Chem 280:3390–3399

Yamamoto K, Yoshida H, Kokame K, Kaufman RJ, Mori K (2004) Differential contributions of ATF6 and XBP1 to the activation of endoplasmic reticulum stress-responsive cis-acting elements ERSE, UPRE and ERSE-II. J Biochem (Tokyo) 136:343–350

Yashiroda H, Oguchi T, Yasuda Y, Toh E, Kikuchi Y (1996) Bul1, a new protein that binds to the Rsp5 ubiquitin ligase in Saccharomyces cerevisiae. Mol Cell Biol 16:3255–3263

Ye Y, Meyer HH, Rapoport TA (2001) The AAA ATPase Cdc48/p97 and its partners transport proteins from the ER into the cytosol. Nature 414:652–656

Ye Y, Meyer HH, Rapoport TA (2003) Function of the p97-Ufd1-Npl4 complex in retrotranslocation from the ER to the cytosol: dual recognition of nonubiquitinated polypeptide segments and polyubiquitin chains. J Cell Biol 162:71–84

Ye Y, Shibata Y, Yun C, Ron D, Rapoport TA (2004) A membrane protein complex mediates retro-translocation from the ER lumen into the cytosol. Nature 429:841–847

Yoshida Y, Chiba T, Tokunaga F, Kawasaki H, Iwai K, Suzuki T, Ito Y, Matsuoka K, Yoshida M, Tanaka K, Tai T (2002) E3 ubiquitin ligase that recognizes sugar chains. Nature 418:438–442

Yoshida Y, Tokunaga F, Chiba T, Iwai K, Tanaka K, Tai T (2003) Fbs2 is a new member of the E3 ubiquitin ligase family that recognizes sugar chains. J Biol Chem 278:43877–43884

Yu H, Kaung G, Kobayashi S, Kopito RR (1997) Cytosolic degradation of T-cell receptor alpha chains by the proteasome. J Biol Chem 272:20800–20804

Yu H, Kopito RR (1999) The role of multiubiquitination in dislocation and degradation of the alpha subunit of the T cell antigen receptor. J Biol Chem 274:36852–36858

Zhang Y, Gao J, Chung KK, Huang H, Dawson VL, Dawson TM (2000) Parkin functions as an E2-dependent ubiquitin- protein ligase and promotes the degradation of the synaptic vesicle-associated protein, CDCrel-1. Proc Natl Acad Sci U S A 97:13354–13359

Zhao J, Ren Y, Jiang Q, Feng J (2003) Parkin is recruited to the centrosome in response to inhibition of proteasomes. J Cell Sci 116:4011–4019

Zheng N, Schulman BA, Song L, Miller JJ, Jeffrey PD, Wang P, Chu C, Koepp DM, Elledge SJ, Pagano M, Conaway RC, Conaway JW, Harper JW, Pavletich NP (2002) Structure of the Cul1-Rbx1-Skp1-F boxSkp2 SCF ubiquitin ligase complex. Nature 416:703–709

Zhong L, Tan Y, Zhou A, Yu Q, Zhou J (2005) RING finger ubiquitin-protein isopeptide ligase Nrdp1/FLRF regulates Parkin stability and activity. J Biol Chem 280:9425–9430

Zhong X, Shen Y, Ballar P, Apostolou A, Agami R, Fang S (2004) AAA ATPase p97/valosin-containing protein interacts with gp78, a ubiquitin ligase for endoplasmic reticulum-associated degradation. J Biol Chem 279:45676–45684

CTMI (2006) 300:95–125

The Role of p97/Cdc48p in Endoplasmic Reticulum-Associated Degradation: From the Immune System to Yeast

S. Bar-Nun (✉)

Department of Biochemistry, George S. Wise Faculty of Life Sciences, Tel Aviv University, Tel Aviv, Israel
shoshbn@tauex.tau.ac.il

Abstract Quality control mechanisms in the endoplasmic reticulum prevent deployment of aberrant or unwanted proteins to distal destinations and target them to degradation by a process known as endoplasmic reticulum-associated degradation, or ERAD. Attempts to characterize ERAD by identifying a specific component have revealed that the most general characteristic of ERAD is that the protein substrates are initially translocated to the ER and eventually eliminated in the cytosol by the ubiquitin-proteasome pathway. Hence, dislocation from the ER back to the cytosol is a hallmark in ERAD and p97/Cdc48p, a cytosolic AAA-ATPase that is essential for ERAD, appears to provide the driving force for this process. Moreover, unlike many ERAD components that participate in degradation of either lumenal or membrane substrates, p97/Cdc48p has a more general role in that it is required for ERAD of both types of substrates. Although p97/Cdc48p is not dedicated exclusively to ERAD, its ability to physically associate with ERAD substrates, with VIMP and with the E3 gp78 suggest that the p97/Cdc48$^{Ufd1/Npl4}$ complex acts as a coordinator that maintains coupling between the different steps in ERAD.

Abbreviations

ALLN	*N*-acetyl-leucyl-leucyl-norlecinal
CFTR	Cystic fibrosis conductance transmembrane regulator
CPY*	Carboxypeptidase Y*
ER	Endoplasmic reticulum
ERAD	ER-associated degradation
Hmg2p	Yeast 3-hydroxy-3-methylglutaryl-coenzyme A reductase
HMG-CoA reductase	Mammalian 3-hydroxy-3-methylglutaryl-coenzyme A reductase
MHC	Major histocompatibility complex
MG-132	Carboxybenzyl-leucyl-leucyl-leucinal
UPR	Unfolded protein response

1 Characteristics of Endoplasmic Reticulum-Associated Degradation

Quality control mechanisms, which operate in the endoplasmic reticulum (ER) and prevent deployment of aberrant or unwanted proteins to distal destinations in the secretory pathway, were recognized in the late 1980s (Klausner and Sitia 1990; Klausner et al. 1990; Hammond and Helenius 1994). It was also evident that such misfolded or unassembled proteins are eventually degraded prior to the Golgi (Chen et al. 1988). However, it was only in 1995 that the ubiquitin-proteasome pathway has been implicated in this degradation, thanks to studies on the intracellular fate of the cystic fibrosis conductance transmembrane regulator (CFTR) (Ward et al. 1995; Jensen et al. 1995). The involvement of the ubiquitin-proteasome pathway, which is located in the cytoplasm and nucleoplasm, indicates that substrates must be dislocated from the ER back to the cytosol in order to be conjugated to ubiquitin and eliminated by the proteasome (Kopito 1997; Riezman 1997; Bonifacino and Weissman 1998). Attempts to characterize this process, known as ER-associated degradation or ERAD, by identifying a specific ERAD component, have revealed that the most general characteristic of ERAD is that the protein substrates are initially translocated to the ER and eventually eliminated in the cytosol by the ubiquitin-proteasome pathway.

1.1 The Proteasome

The role of proteasome as the executing protease in ERAD is well established. The yeast ERAD substrate carboxypeptidase Y* (CPY*) is stabilized in various mutant alleles of the 20S core particle (e.g., *pre1-1*, *pre2-2*, *pre2-K108R*, *pre3-T20A*, *pre4-1*) (Heinemeyer et al. 1991, 1997; Hiller et al. 1996), many mammalian ERAD substrates are stabilized in the presence of a variety of drugs that block the proteolytic activity of the proteasome (Rock et al. 1994; Bogyo et al. 1997), and the 26S proteasome is implicated in the degradation of the yeast 3-hydroxy-3-methylglutaryl-coenzyme A reductase (Hmg2p) (Hampton et al. 1996). The ATPases of the 19S regulatory particle are also involved in ERAD, as shown mostly by studies of yeast mutants (e.g., *cim5-1*, *rpt1S*, *rpt2RF*, *rpt4R*, *rpt5S*, *cim3-1*) (Rubin et al. 1998; Mayer et al. 1998; Jarosch et al. 2002; Hiller et al. 1996; Hill and Cooper 2000). For example, CPY* is stabilized in *rpt4R* and *rpt5S*, but not in *rpt2RF* (Jarosch et al. 2002). However, degradation by the proteasome cannot serve as a hallmark of ERAD, since the proteasome is the major proteolytic system also in the degradation of cytosolic and nuclear proteins. The possibility that distinct proteasome subunits are engaged in ERAD or in cytosolic/nuclear degradation has yet to be explored, especially in light of the findings that a subpopulation of the proteasome is bound to the ER (Rivett 1998; Enenkel et al. 1998; Hori et al. 1999; Brooks et al. 2000; Hirsch and Ploegh 2000; Elkabetz et al. 2004), and the suggested role of the proteasome, including its Cim5/Rpt1p and Rpt4p ATPases, in the extraction of ERAD substrates from the ER (Mayer et al. 1998; Jarosch et al. 2002; Elkabetz et al. 2004) (see below).

1.2
E3 Ubiquitin Ligases

Attempts have been made to characterize ERAD by ER-localized E2 ubiquitin conjugating enzymes, by ER-localized E3 ubiquitin ligases, or by their combinations. However, with the growing number of recognized ERAD substrates, it is becoming evident that, just like with ubiquitination of proteins in general, the combinations of cognate E2s and E3s are determined by the substrate proteins (Cyr et al. 2002). For example, the Hrd1p/Der3p is a transmembrane protein localized to the ER with a cytosol-facing RING-H2 domain and E3 activity that functions in a complex with Hrd3p. For several years, Hrd1p/Der3p-Hrd3p was considered to be the E3 that characterizes ERAD in yeast, because it was shown to be essential for the metabolically regulated degradation of the membrane protein Hmg2p, as well as for the degradation of the lumenal protein CPY* (Hampton et al. 1996; Bordallo et al. 1998). Subsequently, it was shown that ERAD could proceed without Hrd1p/Der3p (Wilhovsky et al. 2000), that the degradation of unassembled Vph1p, a membrane subunit of the yeast V-ATPase, is independent of Hrd1p/Der3p (Hill and Cooper 2000), and even the degradation of CPY* may occur in a HRD/DER-independent HIP pathway, relying instead on the HECT domain E3 Rsp5 (Haynes et al. 2002).

An additional ER-localized transmembrane RING-H2-containing E3 implicated in ERAD is Doa10p/Ssm4p. Although Doa10p has been identified via its role in the degradation of soluble substrates such as Matα2 or any *Deg1*-bearing proteins and is not involved in the ERAD of CPY*, Doa10p is the E3 in the ERAD of Ste6p*, a multispanning membrane protein with a cytosolic mutation, and of Pma1-D387N mutant (Swanson et al. 2001; Huyer et al. 2004; Wang and Chang 2003). Moreover, the severe effect of the *doa10Δ/hrd1Δ* double mutant on ERAD suggests that these two E3s have overlapping functions that include ERAD (Swanson et al. 2001). It appears that Hrd1p/Der3p-Hrd3p is dedicated to soluble lumenal ERAD substrates or to membrane substrates with lumenal recognition motifs, whereas Doa10p is specific for membrane ERAD substrates with cytosolic recognition motifs (Taxis et al. 2003; Huyer et al. 2004; Vashist and Ng 2004). Yet Hrd1p was discovered via stabilization of Hmg2p, a polytopic membrane protein whose recognition motif appears to be scattered throughout its membrane region, including within the transmembrane spans (Hampton et al. 1996; Gardner and Hampton 1999), and the degradation of the polytopic membrane CFTR expressed in yeast requires Hrd1p/Der3p and Doa10p ubiquitin ligases (Gnann et al. 2004).

HRD1, the putative human ortholog of the yeast Hrd1p, is also an ER-resident protein with a cytosolic RING-H2 domain and an E3 activity (Kaneko et al. 2002; Nadav et al. 2003; Kikkert et al. 2004). HRD1 has been implicated in degradation of unfolded proteins that accumulate in the ER (Kaneko et al. 2002), although it is not involved in the sterol-accelerated degradation of the mammalian 3-hydroxy-3-methylglutaryl-coenzyme A reductase (HMG-CoA reductase) (Nadav et al. 2003; Kikkert et al. 2004). On the other hand, HRD1 takes part in the basal degradation of HMG-CoA reductase, as well as in the elimination of TCRα and CD3-δ, two model substrates for ERAD in mammalian cells (Kikkert et al. 2004). Interestingly, CD3-δ is also recognized as the substrate of gp78 (also known as tumor autocrine motility factor receptor), another ER-embedded RING E3 in mammalian cells (Fang et al. 2001). Additional mammalian E3s that were implicated in ERAD are Parkin and the F-box proteins Fbx2/Fbs1 and Fbs2. Parkin is the E3 of the Pael receptor, another ERAD substrate (Imai et al. 2001) and Fbx2 and Fbs2 recognize N-glycans via their F-box motifs and participate in ubiquitination of glycoproteins in general, and in ERAD of TCRα in particular (Yoshida et al. 2002, 2003). CHIP, a chaperone-dependent ubiquitin protein ligase, represents the U-box proteins, another class of E3 enzymes that may contribute to ERAD. CHIP binds Hsc70 and Hsc90 via its N-terminal TRP domain and turns these molecular chaperones into protein degradation factors. Through its C-terminal U-box, a modified form of the RING domain, CHIP promotes polyubiquitination of chaperone-bound substrates (e.g., CFTR) and targets them to proteasomal degradation (Meacham et al. 2001; Cyr et al. 2002). In addition, CHIP regulates Parkin activity by facilitating the Parkin-mediated ubiquitination of the Pael receptor (Imai et al. 2002). Finally, the recently discovered co-chaperone HspBP1 attenuates the ubiquitin ligase activity of CHIP when complexed with Hsc70 and consequently interferes with the CHIP-induced degradation of CFTR (Alberti et al. 2004). Hence, it appears that ERAD cannot be defined by a specific E3, since several E3s may recognize the same ERAD substrate and different ERAD substrates are recognized by distinct E3s.

1.3
E2 Ubiquitin Conjugating Enzymes

Defining ERAD by the E2 ubiquitin conjugating enzymes is no less confusing. Initially, Ubc6p and Ubc7p were implicated in yeast as the E2s responsible for ERAD of many substrates, including CPY*, Hmg2p, Sec61-2p, Ste6-166p, Pdr5*, and Fur-430Np (Biederer et al. 1996; Hiller et al. 1996; Hampton and Bhakta 1997; Galan et al. 1998; Loayza et al. 1998; Plemper et al. 1998). This finding was especially attractive because Ubc6p is anchored in the ER membrane and Ubc7p is recruited to the ER membrane by Cue1p (Sommer and Jentsch 1993; Biederer et al. 1997). Furthermore, Ubc6p and Ubc7p were shown to collaborate with the two major E3s implicated in ERAD, Hrd1p/Der3p (Bays et al. 2001a) and Doa10p (Swanson et al. 2001). However, it is now evident that Ubc6p plays only a minor role, if any, in ERAD, whereas the major contributors to ERAD are the soluble Ubc1p and the ER recruit Ubc7p (Friedlander et al. 2000; Bays et al. 2001a). However, the Hrd1p/Der3p-independent degradation of the unassembled Vph1p is also not affected in a *ubc6Δubc7Δ* double mutant (Hill and Cooper 2000). Moreover, as discussed above, the degradation of CPY* by the HRD/DER-independent HIP pathway involves Rsp5 (Haynes et al. 2002), a HECT domain E3 that appears to collaborate with the Ubc4/Ubc5 subfamily of E2s (Gitan and Eide 2000). Interestingly, Ubc4p and possibly Ubc1p and Ubc2p are implicated in the degradation of the unassembled Vph1p, since this substrate is stabilized in *ubc1 ubc4* or *ubc2 ubc4* double mutants (Hill and Cooper 2000).

MmUBC6 and MmUBC7, the mammalian homologs of Ubc6 and Ubc7, are E2 enzymes that form thiol ester bond with ubiquitin in the presence of E1 (Tiwari and Weissman 2001). Again, MmUBC6 is a transmembrane protein anchored to the ER membrane, and although MmUBC7 is not an integral membrane protein, it is also localized to the ER (Tiwari and Weissman 2001). Involvement in ERAD has been shown only for MmUBC7, where its C89S mutant acts in a dominant negative fashion and delays the degradation of TCRα and CD3-δ (Tiwari and Weissman 2001). Finally, the question of whether

ERAD can be defined by specific E2 or E3 or a set of cognate E2s and E3s has been further complicated by a recent paper describing two pathways for the degradation of CD4 in yeast (Meusser and Sommer 2004). In HIV-infected T cells, CD4 elimination by ERAD is triggered by the HIV-encoded Vpu, whose phosphorylated form is recognized by the F-box E3 βTrCP. When expressed in yeast, CD4 is eliminated by the cellular ERAD, involving Ubc7p, Ubc1p, Hrd1p, and Hrd3p. However, in *hrd3Δ* yeast strain, CD4 is stabilized unless Vpu and βTrCP are also co-expressed. Under these conditions, the degradation of CD4 no longer depends on Ubc7p or Hrd1p (Meusser and Sommer 2004).

1.4
The Vesicular Transport Dependence or Different Routes to ERAD for Lumenal and Membrane Substrates

Another complexity in the definition of the ERAD pathway is whether the substrate must exit the ER by vesicular transport before it dislocates to the cytosol, since one of the hallmarks of ER protein degradation in the early studies was that it occurs regardless of vesicular traffic (Stafford and Bonifacino 1991). In our study in B cells on the degradation of μs, the heavy chain of the secretory IgM (sIgM), we were the first to describe that the vesicular exit of this lumenal substrate from the ER is a prerequisite for its degradation. Our biochemical data were based on inhibition of vesicular transport by a temperature block, by drugs such as brefeldin A or by cell permeabilization (Amitay et al. 1991, 1992; Rabinovich et al. 1993; Winitz et al. 1996). Recently, genetic data in yeast have indicated that genes involved in ER-to-Golgi vesicular transport, such as *SEC12*, *SEC18*, and *ERV29*, are required for ERAD of lumenal substrates such as CPY* and PrA*, but not for ERAD of membrane substrates (Caldwell et al. 2001; Vashist et al. 2001; Haynes et al. 2002; Spear and Ng 2003). A very recent report extends the requirement for vesicular transport to membrane ERAD substrates with lumenal recognition motifs (Vashist and Ng 2004). Although the purpose of this vesicular journey is not yet clear, it adds a level of complexity to the definition of ERAD. In fact, the studies in yeast suggest that the vesicular transport is essential only for getting rid of excess CPY*, above the levels that saturate the vesicular transport-independent ERAD (Haynes et al. 2002; Spear and Ng 2003). Our data in B-lymphocytes suggest that the vesicular transport is related to the differentiation of pre-B cells, which do not express any conventional Ig light chain, to light chain-expressing B-cells. We have shown that transport blockers attenuate μs degradation only when this Ig heavy chain is assembled with conventional light chains (Elkabetz et al. 2003) (see Fig. 1). Nonetheless, both pathways of μs degradation converge at the ubiquitin-proteasome pathway (Elkabetz et al. 2003) (see Fig. 2).

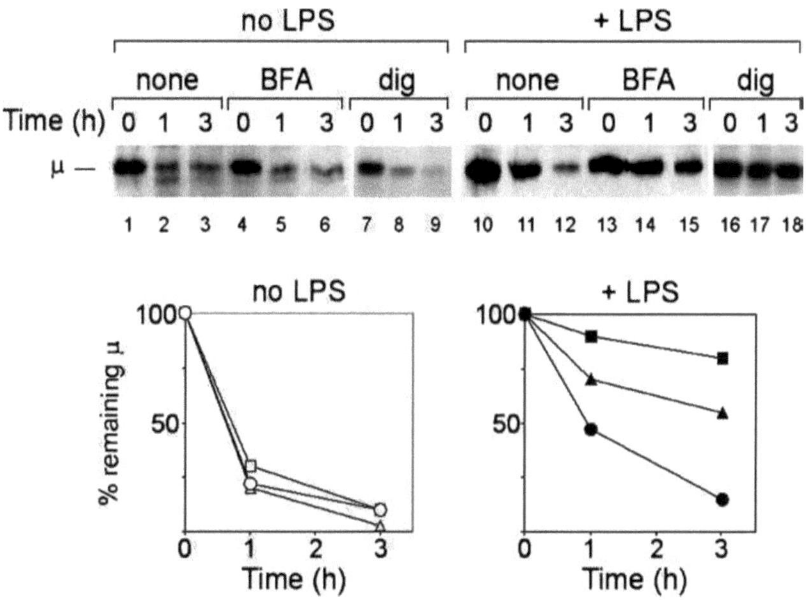

Fig. 1 Degradation of μ heavy chains depends on vesicular transport only upon expression of κ light chain. 70Z/3 pre-B cells stimulated (18 h) with (+ LPS, *closed symbols*) or without (no LPS, *open symbols*) lipopolysaccharide were pulse-labeled (10 min) with ^{35}S-methionine and chased for the indicated time either in vivo as intact cells with (BFA, *squares*) or without (none, *circles*) brefeldin A, or in vitro following permeabilization with digitonin (dig, *triangles*). Immunoprecipitated IgM was resolved by reducing SDS-PAGE and detected by autoradiography. *μ*, μ heavy chains. ^{35}S-labeled μ was quantified by densitometry (five independent in vivo experiments and two independent in vitro experiments) and remaining μ was calculated as a percentage of its level at the end of the pulse (100%). (From Elkabetz et al. 2003; courtesy of *The Journal of Biological Chemistry*)

1.5 The Translocon

Since all ERAD substrates are initially translocated into the ER through a proteinaceous channel, the Sec61 translocon, and are eventually eliminated in the cytosol by the ubiquitin-proteasome pathway, dislocation is a hallmark of ERAD. Yet, the putative channel through which ERAD substrates are dislocated back to the cytosol, the dislocon, has not been fully defined. Biochemical evidence in mammalian cells (Wiertz et al. 1996) and genetic data in yeast (Pilon et al. 1997; Plemper et al. 1997; Zhou and Schekman 1999; Wilkinson et al. 2000; reviewed in Romisch 1999) suggest that the Sec61 translocon also functions as the dislocon. However, the ERAD of Ste6p*, unlike CPY*

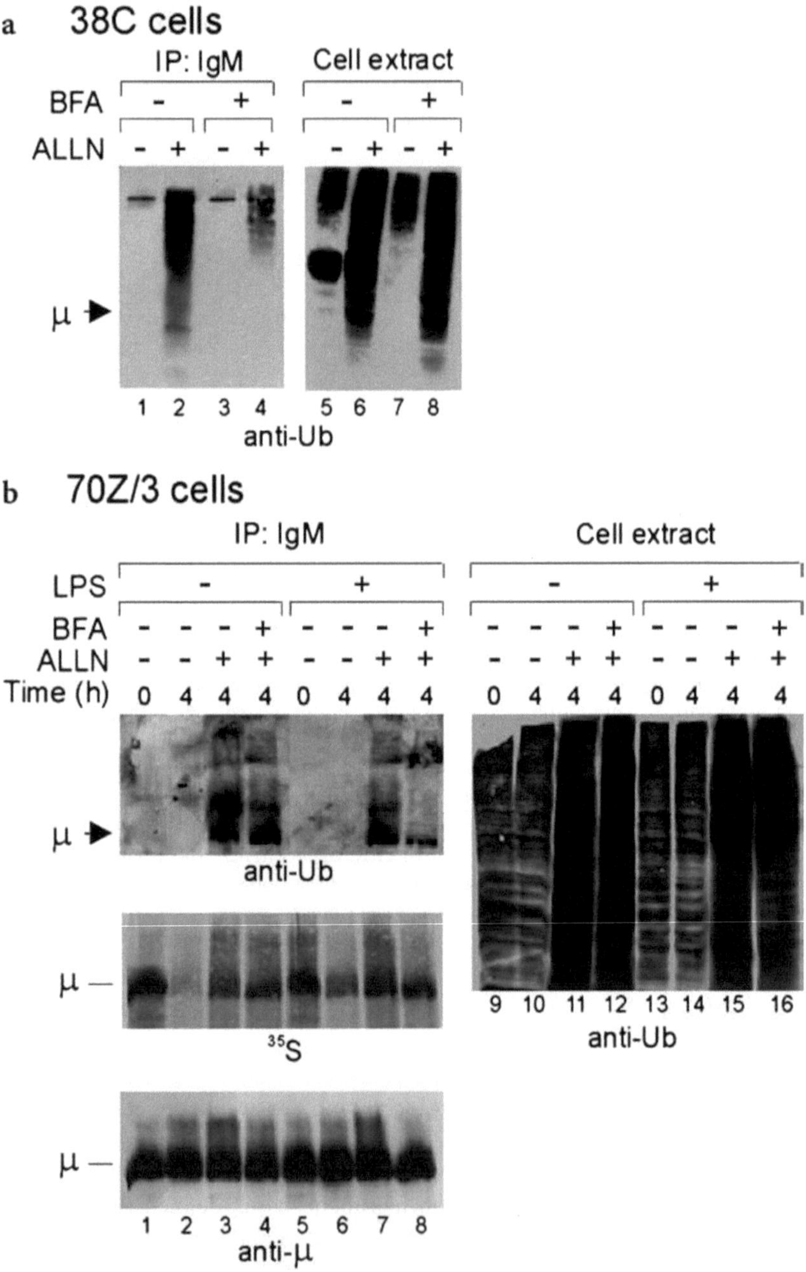
a
38C cells
IP: IgM
Cell extract
BFA
ALLN
μ
1 2 3 4
5 6 7 8
anti-Ub
b
70Z/3 cells
IP: IgM
Cell extract
LPS
BFA
ALLN
Time (h)
0 4 4 4 0 4 4 4
0 4 4 4 0 4 4 4
μ
anti-Ub
μ
35S
μ
1 2 3 4 5 6 7 8
anti-μ
9 10 11 12 13 14 15 16
anti-Ub

Fig. 2a, b Ubiquitination of μ is detected upon proteasome inhibition in pre-B and B cells, yet it is inhibited by brefeldin A only in B cells. 38C B cells (**a**) or 70Z/3 pre-B cells stimulated (18 h) with (+) or without (–) lipopolysaccharide (*LPS*) and pulse-labeled (5 min) with ^{35}S-methionine (**b**), were either incubated or chased (4 h) with (+) or without (–) proteasome inhibitor (*ALLN*), transport blocker [brefeldin A (*BFA*)], or both. Immunoprecipitated IgM (IP: IgM; **a**, lanes *1–4*; **b**, lanes *1–8*), and total proteins from cell extracts (**a**, 1%, lanes *5–8*; **b**, 10%, lanes *9–16*), were resolved by reducing SDS-PAGE, electroblotted, and the blot was autoradiographed (^{35}S; **b**, lanes *1–8*, *middle panel*) and then probed with a mouse antiubiquitin antibody followed by a horseradish peroxidase (HRP)-conjugated anti-mouse antibody (*anti-Ub*; **a**; **b**, lanes *1–8*, *upper panel*) and reprobed with an HRP-conjugated anti-μ antibody (*anti-μ*; **b**, lanes *1–8*, *lower panel*) and the HRP was visualized by enhanced chemiluminescence (ECL). *Arrow*, the migration of μ heavy chains, as detected by an anti-μ antibody. (From Elkabetz et al. 2003; courtesy of *The Journal of Biological Chemistry*).

degradation, does not appear to involve the Sec61p, but similar to CPY*, may depend on the Sec61p homolog Ssh1p (Huyer et al. 2004). Recent data implicate Derlin-1, the mammalian homolog of the yeast Der1 (Knop et al. 1996), as a candidate to play a role in dislocation (Ye et al. 2004; Lilley and Ploegh 2004). Derlin-1 associates with different ERAD substrates, with p97 (see below), and with US11, a human cytomegalovirus protein that targets the major histocompatibility complex (MHC) class I heavy chain to ERAD (Ye et al. 2004; Lilley and Ploegh 2004). Moreover, Derlin-1 is required for the US11-catalyzed but not for the US2 mediated degradation of MHC class I heavy chain, and inactivation of Derlin-1 in *Caenorhabditis elegans* causes ER stress (Ye et al. 2004; Lilley and Ploegh 2004). However, although inactivation of Der1 in yeast stabilizes ERAD substrates, it appears that *DER1* is essential only for ERAD of soluble lumenal substrates but dispensable for degradation of membrane substrates, even if they carry a lumenal recognition motif (Taxis et al. 2003; Vashist and Ng 2004). Finally, Doa10p is another attractive candidate to play a role in dislocation, because this E3 is anchored in the ER membrane via 14 predicted transmembrane segments (Swanson et al. 2001). Regardless of the actual nature of the dislocon, the dislocation is a fundamental process in ERAD and as such, chaperones at both sides of the ER membrane should provide the driving force to push or pull the substrate back to the cytosol.

1.6
Lumenal Chaperones

Lumenal chaperones are predicted to play a role in the delivery of substrates to the dislocon. The lumenal soluble PDI (protein disulfide isomerase) appears to play a key role in this process (Gillece et al. 1999; Tsai et al. 2001). The calnexin cycle, which relies on reglucosylation by UDP-glucose glycoprotein glucosyltransferase, also plays an important role in ERAD (Hebert et al. 1995; Ellgaard and Helenius 2003). However, in yeast there is no equivalent of glucosyltransferase and although *CNX* is required for the degradation of the nonglycosylated pro-α-factor, this ER lectin appears dispensable for the ERAD of CPY* (McCracken and Brodsky 1996). The distinction between membrane and lumenal ERAD substrates, which is illustrated above with respect to vesicular transport and the E3s, can be extended to chaperones that participate in ERAD. Membrane ERAD substrates may be recognized through motifs located

Fig. 3a–g Ubiquitinated μs accumulates at the cytosolic face of ER in association with microsome-bound proteasome and p97. 38C B cells incubated without (none) or with proteasome inhibitors (ALLN, MG [MG-132; carboxybenzyl-leucyl-leucyl-leucinal]), were disrupted, treated with (+) or without (–) trypsin, and P10 (*P*) and S200 (*S*) fractions were separated by centrifugation. Immunoprecipitated IgM (IP: anti-μ; **a**) or total lysates (**c**, 20%) were resolved by reducing SDS-PAGE and electroblotted proteins were probed (*IB*) and reprobed with the indicated antibodies: anti-μ, anti-BiP, antiubiquitin (*anti-Ub*), anti-p97, anti-proteasome α subunit (*anti*-α), conjugated to HRP or followed by HRP-conjugated secondary antibodies. **b** Immunoprecipitates (*IP*) obtained in parallel with anti-μ or anti-BiP from cells or fractions were resolved by nonreducing (*upper panel*) or reducing (*lower panel*) SDS-PAGE and electroblotted proteins were probed with an HRP-conjugated anti-μ antibody (IB: anti-μ). **d** μs (**a**, *upper panel*) was quantified by densitometry in the various fractions [cytosolic, lumenal (trypsin-resistant) or dislocated (trypsin-sensitive)] and calculated as a percentage of its sum (100%) in the various fractions in each treatment (none, ALLN, MG). The data are the mean of three independent experiments. **e** Microsomal μs was calculated as a percentage of its level in microsomes from cells incubated with no inhibitors that were not treated with trypsin (100%). The data are the mean of three independent experiments. **f** P10 microsomes were resuspended in 0.5 M KCl and wash supernatant and washed microsomes were separated by centrifugation, proteins were resolved by reducing SDS-PAGE, electroblotted, probed with an anti-μ antibody (IB: anti-μ; *upper panel*) and reprobed with an anti-BiP antibody (IB: anti-BiP; *lower panel*). **g** Levels of BiP, proteasome α subunit and p97 that co-precipitated with μ from microsomes of untreated (none) or proteasome-inhibited cells (ALLN, MG-132) that were treated with (+) or without (–) trypsin (*all panels* in **a**), were quantified by densitometry. The values represent the fold increase above untreated cells, which were set as 1. The data are the mean of three independent experiments. The HRP was visualized by ECL. *μ*, free μ heavy chain; *μκ*, hemimers; $\mu_2\kappa_2$, monomers; *μs*, secretory μ; *μm*, membrane μ; *Ub-μ*, ubiquitinated μ. (From Elkabetz et al. 2004; courtesy of *The Journal of Biological Chemistry*) in the cytosol, in the ER lumen, or even within the membrane spans, whereas lumenal ERAD substrates must be recognized only in the ER lumen. Therefore, it is not surprising that distinct sets of chaperones are implicated in ERAD of lumenal or membrane substrates. For example, the major lumenal Hsp70 Kar2p/BiP is required for dislocation and degradation of the lumenal ERAD substrates CPY* and the variant α1PiZ of α1 antitrypsin when expressed in yeast, while this chaperone is dispensable for the degradation of membrane ERAD substrates such as recombinant type I membrane proteins carrying CPY* as a lumenal recognition motif, Ste6p*, unassembled Vph1p and CFTR when expressed in yeast (McCracken and Brodsky 1996; Plemper et al. 1997; Taxis et al. 2003; Huyer et al. 2004; Brodsky et al. 1999; Hill and Cooper 2000).

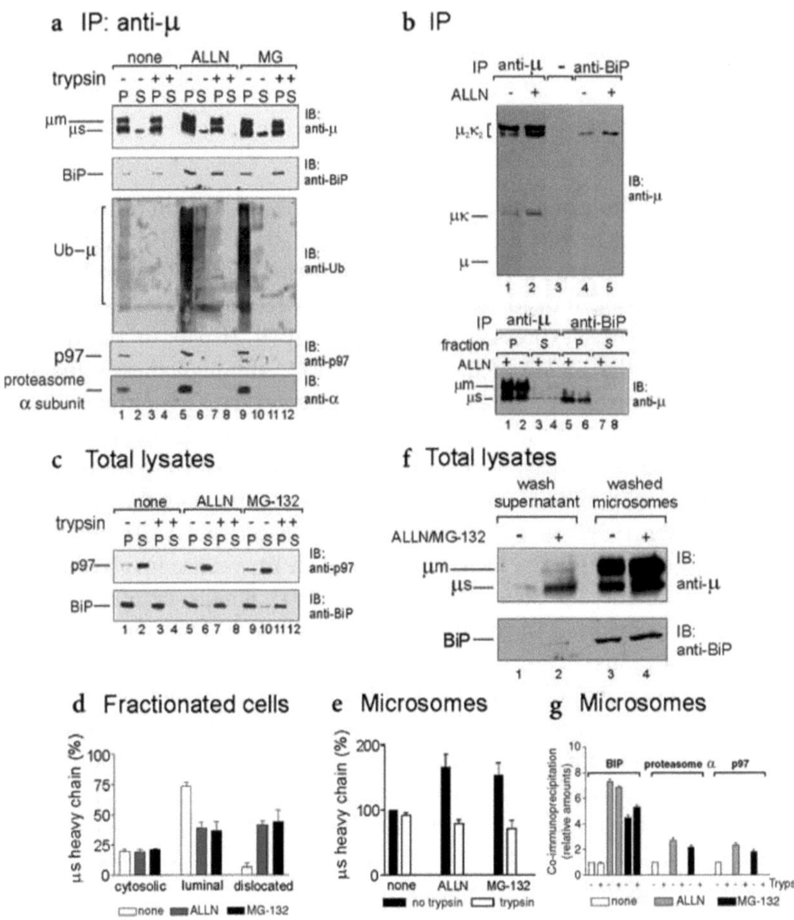

Likewise, the ER lumenal DnaJ-like proteins Jem1p and Scj1p collaborate with Kar2p in preventing aggregation and facilitating degradation of lumenal substrates such as CPY*, yet they are not involved in the ERAD of membrane substrates such as Sec61-2p, the temperature-sensitive unstable mutant form of Sec61p (Nishikawa et al. 2001). Association of BiP with mammalian ERAD substrate has been reported for the pro-parathyroid hormone-related peptide (Meerovitch et al. 1998). Our results in mammalian B cells indicates that in addition to the well-established role of BiP in Ig assembly (Hendershot 1990; Lee et al. 1999), this lumenal molecular chaperone also participates in the ERAD of the lumenal μs heavy chain (Elkabetz et al. 2004). Although BiP is displaced from its stable binding to Ig heavy chains by conventional light chains (Hendershot 1990; Lee et al. 1999), in B cells that express light chains in abundance, BiP still binds μs, although this heavy chain is already assembled with conventional light chains. Under these conditions, BiP does not bind μm, the stable heavy chain of the membrane form of IgM (Elkabetz et al. 2004) (see Fig. 3b).

1.7
Cytosolic Chaperones

By analogy to the role of lumenal ATPases such as BiP/Kar2p in translocation of nascent proteins to the ER (Vogel et al. 1990; Matlack et al. 1999), cytosolic ATPases are predicted to provide the driving force for pulling ERAD substrates back to the cytosol. Indeed, cytosolic Hsp70 is required for the ERAD of the membrane CFTR expressed in yeast, yet Hsp70, Hsp40, or Hsp104 are dispensable for the ERAD of the lumenal CPY*, although these cytosolic chaperones are involved in the degradation of two type I membrane ERAD substrates carrying CPY* as a lumenal recognition motif (Zhang et al. 2001; Taxis et al. 2003). Similarly, the major cytosolic Hsp70 Ssa1p is not required for the degradation of the lumenal α1PiZ variant expressed in yeast (Brodsky and McCracken 1999). Ssa1p and the Hsp40 co-chaperones Ydj1p and Hlj1p are also dispensable for the ERAD of lumenal CPY*, yet are involved in the ERAD of the membrane Ste6p* (Huyer et al. 2004). Finally, cytosolic Hsp70 and Hsp90 are involved in the in vivo and in vitro degradation of human apoprotein B48 (Gusarova et al. 2001). A remarkable addition to the repertoire of cytosolic ATPases that play a role in ERAD has been recently contributed by several groups, who discovered that the cytosolic p97/Cdc48p is an essential ERAD component both in yeast and in mammalian cells (Bays et al. 2001b; Ye et al. 2001; Rabinovich et al. 2002; Jarosch et al. 2002; Braun et al. 2002). This AAA-ATPase has multiple roles and in order to function in ERAD, it collaborates with Ufd1p and Npl4p and functions as $p97/Cdc48^{Ufd1/Npl4}$ complex (Bays et al. 2001b; Ye et al. 2001; Jarosch et al. 2002; Braun et al. 2002).

2
The p97/Cdc48p and ERAD

2.1
The Secretory IgM and its μs Heavy Chain as a Model Mammalian Lumenal ERAD Substrate

We showed in the early 1990s that the secretory form of IgM (sIgM) is rapidly degraded in B cells (Amitay et al. 1991). The degraded μs heavy chains are already assembled with the conventional light chain κ, but do not form the secreted polymers (Shachar et al. 1992). Biochemical analyses have revealed that the degradation of μs is nonlysosomal and occurs prior to the *trans*-Golgi (Amitay et al. 1991). Hence, it appeared to comply with the characteristics of ER degradation as it was known in the early 1990s (Klausner and Sitia 1990; Klausner et al. 1990; Hammond and Helenius 1994). However, μs is unique because it is a soluble lumenal protein, while most substrates of ER degradation that were studied in mammalian cells were membrane proteins. Indeed, already in 1992 we showed that the degradation of μs requires vesicular transport, since it is attenuated by brefeldin A or in permeabilized cells (Amitay et al. 1992; Winitz et al. 1996). Moreover, the degradation of μs cannot be considered as an elimination of an aberrant protein, since the same μs molecule is efficiently secreted from plasma cells (Amitay et al. 1991; Shachar et al. 1992). Nonetheless, the first indication that μs is degraded by the proteasome was provided already in 1992, when μs was stabilized in the presence of *N*-acetyl-leucyl-leucyl-norlecinal (ALLN), then considered as an inhibitor of calpain or of cysteine proteases (Amitay et al. 1992). Subsequently, we showed that μs is indeed a bona fide lumenal ERAD substrate, which is handled by the ubiquitin-proteasome pathway (Elkabetz et al. 2003) (see Fig. 2).

As discussed above, recent genetic findings in yeast (Caldwell et al. 2001; Vashist et al. 2001; Haynes et al. 2002; Spear and Ng 2003; Vashist and Ng 2004) corroborate our initial biochemical findings in B cells (Amitay et al. 1992), demonstrating that ERAD of soluble lumenal substrates depends on vesicular transport. Nonetheless, unlike the suggestion that vesicular transport is required only to handle an excess of CPY* (Haynes et al. 2002; Spear and Ng 2003), in the case of μs the vesicular transport is tightly linked to its assembly status (Rabinovich et al. 1993; Winitz et al. 1996), which is correlated with the differentiation stage of the B lymphocytes (Elkabetz et al. 2003). We showed that assembly with conventional light chains diverts μs degradation from a vesicular transport-independent to a vesicular transport-dependent process (Elkabetz et al. 2003) (see Fig. 1). Nonetheless, both routes comply with the characteristics of ERAD, as they converge at the ubiquitin-proteasome pathway (Elkabetz et al. 2003) (see Fig. 2). Interestingly, in the case of μs, the vesicular transport blocker brefeldin A also attenuates the ubiquitination of μs, but only upon assembly with light chains, indicating that the vesicular transport is a prerequisite for ubiquitination (Elkabetz et al. 2003) (see Fig. 2).

2.2
The Discovery of p97/Cdc48p as an Essential Component in ERAD

Our study on the ERAD substrate μs in B cells, in comparison with the stable and secreted μs in plasma cells, allows us to search for components that play a specific role in ERAD. One such example is the ER lumenal chaperone BiP. As discussed above, despite its well-established role in Ig assembly, BiP is found in association only with the ERAD substrate μs, but not with the stable μm, and this interaction is not counteracted by the excess light chains in B cells (Elkabetz et al. 2004) (see Fig. 3b). Moreover, because μs is a lumenal ERAD substrate, it enables us to look for cytosolic factors that interact specifically with the dislocated μs, subsequent to its emergence from the ER. Therefore, we were very excited when we co-precipitated p97 with μs from B cells, but not from plasma cells, although p97 was equally abundant in both cell types and μs was much more abundant in the plasma cells (Rabinovich et al. 2002) (see Fig. 4). Consequently, in collaboration with Dr. Kai-Uwe Fröhlich from Graz university, we unequivocally demonstrated in yeast *cdc48* conditional mutants (*cdc48–1, cdc48–10*) the stabilization of two well-established ERAD

Fig. 4a–c p97/VCP is co-precipitated with unstable μs in B-cells. **a** 38C B cells were lysed in the absence (–) or presence (+) of cross-linking reagent (disuccinimidyl suberate, *DSS*). IgM and IgM-containing cross-linked complexes were immunoprecipitated with goat anti-IgM antibodies (IP: anti-IgM), resolved by SDS-PAGE, electroblotted and probed with a rabbit anti-μs antibody followed by HRP-conjugated anti-rabbit IgG (IB: anti-μs). The precipitated μs and ~100 kDa μs-complex are indicated. The band of ~100 kDa, marked by the *asterisk*, was subjected to MALDI-mass spectrometry and was identified to be p97. **b** IgM and proteins complexed with IgM were immunoprecipitated from 38C B cells with goat anti-IgM antibodies (IP: anti-IgM). Immunoprecipitates and total cell extract (5%) were resolved by SDS-PAGE, electroblotted and probed with a mouse anti-p97 antibody followed by HRP-conjugated anti-mouse IgG (IB: anti-p97). The pulled-down p97 is indicated. **c** Total cell extracts (10%, *right panel*) and μs and μs-associated proteins that precipitated with anti-IgM antibodies (IP: anti-IgM, *left panels*), from 38C B cells, D2 plasma cells and U2OS nonlymphoid cells, were resolved by SDS-PAGE and electroblotted. Upper blots were probed with mouse anti-p97 followed by HRP-conjugated anti-mouse IgG (IB: anti-p97) and the blot in the *left panel* was reprobed with rabbit anti-μs followed by HRP-conjugated anti-rabbit IgG (IB: anti-μs). HRP was visualized by ECL. (From Rabinovich et al. 2002; courtesy of *Molecular and Cellular Biology*) substrates, membrane 6myc-Hmg2p (see Fig. 6) and lumenal CPY* (see Fig. 5) (Rabinovich et al. 2002). Moreover, we have shown that 6myc-Hmg2p physically interacts with Cdc48p (Rabinovich et al. 2002).

The discovery of p97/Cdc48p and its partners Ufd1p and Npl4p as essential components in ERAD has been shared by five groups. Bays et al. identified *NPL4* as *HRD4* (Bays et al. 2001b). Ye et al. showed that CPY* as well as MHC class I heavy chain expressed in yeast are stabilized in strains that harbor mutant alleles of Cdc48p, Ufd1p, or Npl4p (*cdc48-1, cdc48-3, ufd1-1, npl4-1*) (Ye et al. 2001). This group also showed that mammalian p97 and its ATPase activity are required to facilitate the in vitro release of MHC class I heavy chain from microsomes of US11-expressing cells, and that this in vitro process is

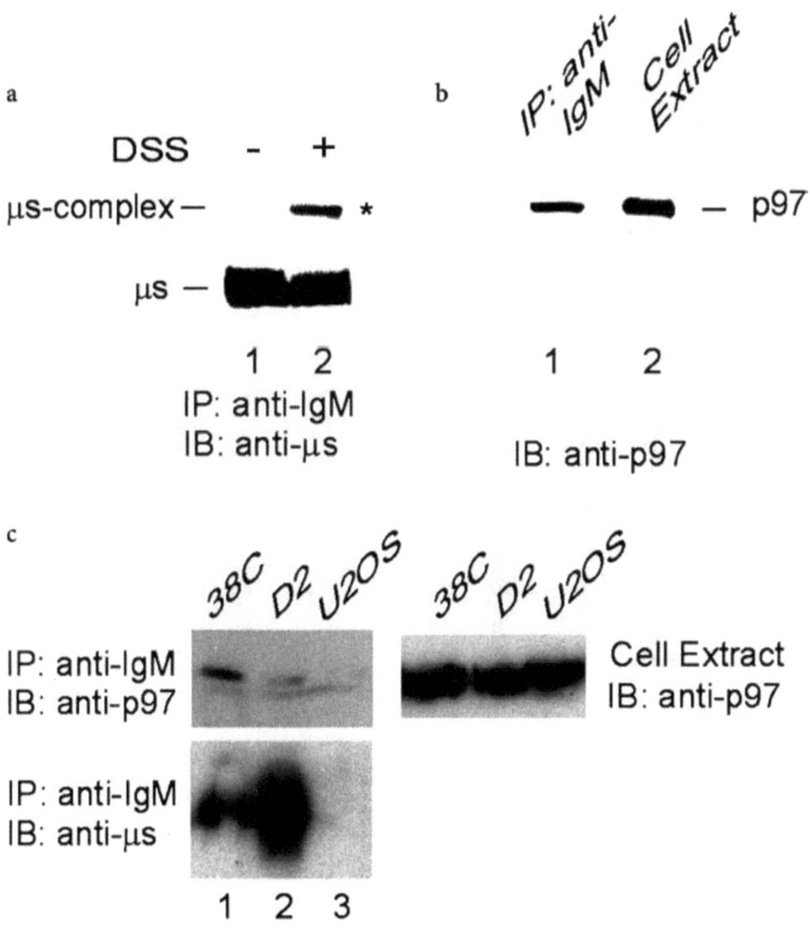

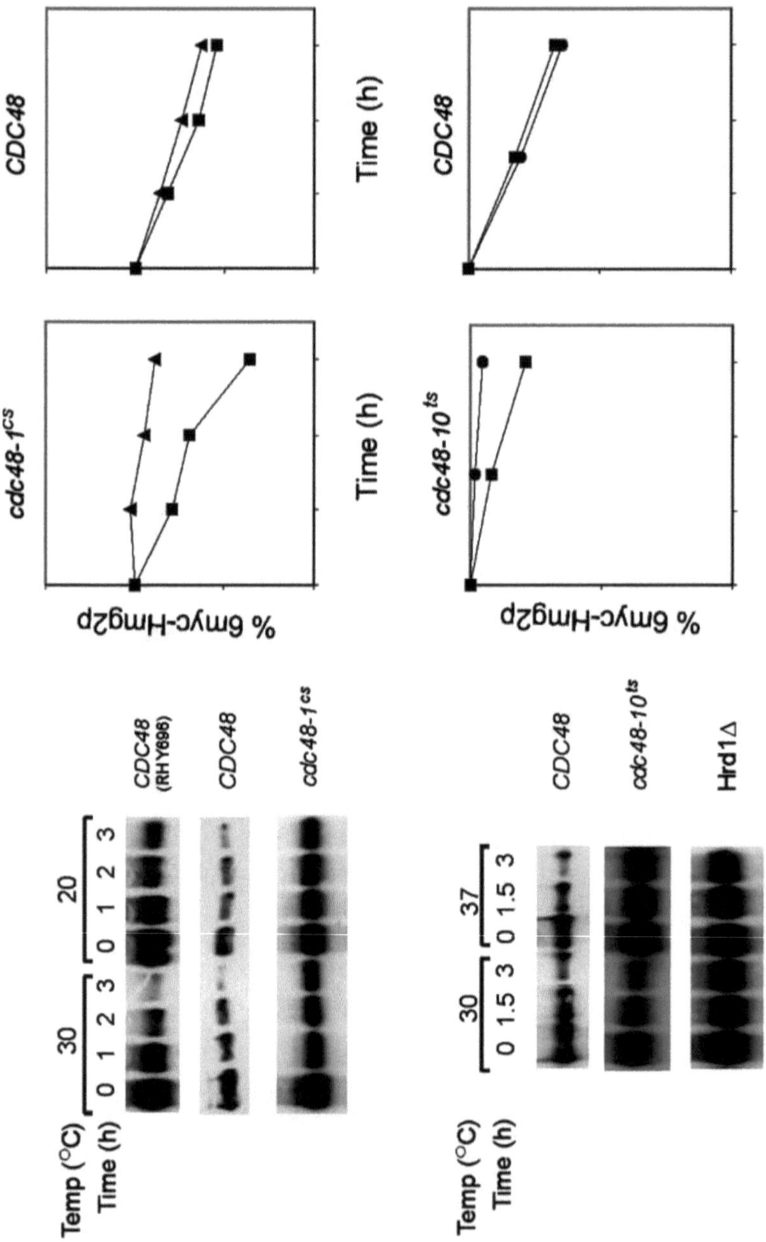
CDC48
cdc48-1cs
% 6myc-Hmg2p
Time (h)
cdc48-10ts
Temp (°C)
Time (h)
30
20
0 1 2 3
CDC48 (RHY696)
cdc48-1cs
37
0 1.5 3
Hrd1Δ

Fig. 5 The ERAD of 6myc-Hmg2p is impaired in yeast cdc48 mutants. Stability of 6myc-Hmg2p was measured in strains KFY100 (wild type *CDC48*), KFY116 (*cdc48-1*^{cs} mutation) and KFY194 (*cdc48-10*^{ts} mutation) expressing 6myc-Hmg2p (plasmid pRH244). Strains RHY696 and RHY965 express integrated 6myc-Hmg2p and carry wild type *CDC48*, and strain RHY965 is also *hrd1Δ*. Following 2 h preincubation at the indicated permissive or restrictive temperatures, cycloheximide (100 μg/ml) was added and cells were collected at the indicated time points. Total cellular proteins were resolved by reducing SDS-PAGE, electroblotted, probed with a mouse anti-myc antibody followed by HRP-conjugated anti-mouse IgG and the HRP was visualized by ECL. The blots were quantified and the remaining 6myc-Hmg2p, calculated as a percentage of its level at the time of cycloheximide addition (100%), is plotted in semi-logarithmic scale representing its decay. *Upper panels*, *cdc48-1*^{cs} (*left*) and wild type *CDC48* (*right*) at 20°C (△) or 30°C (□). *Lower panels*, *cdc48-10*^{ts} (*left*) and wild type *CDC48* (*right*) at 37°C (○) or 30°C (□). (From Rabinovich et al. 2002; courtesy of *Molecular and Cellular Biology*)

inhibited by an excess of p47, the partner that competes with Ufd1p and Npl4p for binding to p97 (Ye et al. 2001). Using the same *cdc48-1*, *ufd1-1*, and *npl4-1* alleles, as well as expression of *cdc48*^{Y834A} or *cdc48*^{S565G} plasmids on the background of genomic *CDC48* deletion, Jarosch et al. showed that the p97/Cdc48$^{Ufd1/Npl4}$ complex is essential for the ERAD of CPY* (Jarosch et al. 2002). Braun et al. showed that *CDC48(UFD1/NPL4)* are required for the ERAD of OLE1 (Braun et al. 2002). Since then, the p97/Cdc48$^{Ufd1/Npl4}$ has been found to be essential for ERAD of every substrate studied in yeast (e.g., Wang and Chang 2003; Taxis et al. 2003; Medicherla et al. 2004; Gnann et al. 2004). Hence, unlike many ERAD components that participate in degradation of either lumenal or membrane substrates, p97/Cdc48p appears to be a common denominator with a more general role, since it is required for ERAD of both types of substrates.

The involvement of p97/Cdc48p in proteasomal degradation was already reported in 1996, when the yeast *CDC48* was implicated in the degradation of UFD substrates, but not of N-end rule substrates, although both are substrates of the proteasome (Ghislain et al. 1996). In 1998, the mammalian p97 was detected in a complex with the 26S proteasome and with IκBα, a cytosolic substrate of the proteasome (Dai et al. 1998). Subsequently, it was shown that p97 interacts directly with polyubiquitin chains, preferentially with a minimal length of 4 ubiquitin units (Dai and Li 2001; Rape et al. 2001). Although this interaction is relatively weak (Dai and Li 2001; Rape et al. 2001), it is proposed to be assisted by the Ufd1p–Npl4p complex, which by itself can bind polyubiquitin via the Npl4p zinc finger domain (Meyer et al. 2002; Wang et al. 2003). Interestingly, p97 in the context of p97/p47 complex binds monoubiq-

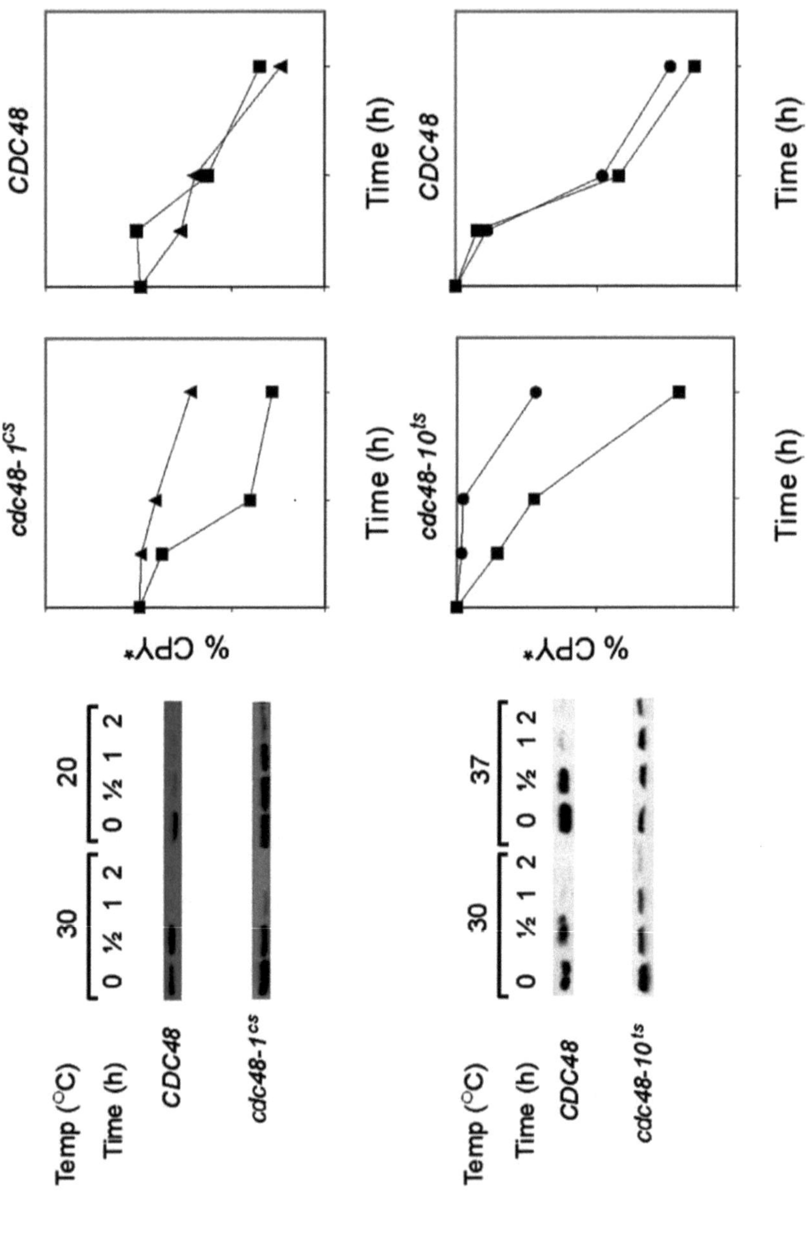
CDC48
cdc48-1cs
cdc48-10ts
Time (h)
% CPY*
Temp (°C)
30
20
37
0 ½ 1 2

Fig. 6 ERAD of CPY* is impaired in yeast *cdc48* mutants. Stability of CPY* was measured for integrated CPY* expressed in strains KFY100 (wild type *CDC48)*, KFY116 (*cdc48-1^{cs}* mutation) and KFY194 (*cdc48-10^{ts}* mutation). Following 2 h preincubation at the indicated permissive or restrictive temperatures, cycloheximide (100 μg/ml) was added and cells were collected at the indicated time points. Total cellular proteins were resolved by reducing SDS-PAGE, electroblotted, probed with a mouse anti-CPY antibody followed by HRP-conjugated anti-mouse IgG and the HRP was visualized by ECL. The blots were quantified and remaining CPY*, calculated as a percentage of its level at the time of cycloheximide addition (100%), and plotted in semi-logarithmic scale. *Upper panels, cdc48-1^{cs}* (*left*) and wild type *CDC48* (*right*) at 20°C (△) or 30°C (□). *Lower panels, cdc48-10^{ts}* (*left*) and wild type *CDC48* (*right*) at 37°C (○) or 30°C (□). (From Rabinovich et al. 2002; courtesy of *Molecular and Cellular Biology*)

uitin rather than polyubiquitin (Meyer et al. 2002). Together these findings suggest that p97/Cdc48p recognizes polyubiquitinated substrates and present them to the proteasome. The physical interaction of p97/Cdc48$^{Ufd1/Npl4}$ with the proteasome on one hand (Dai et al. 1998; Verma et al. 2000) and with polyubiquitinated ERAD substrates on the other hand (Ye et al. 2001; Rabinovich et al. 2002; Elkabetz et al. 2004) may assign p97 the responsibility for targeting polyubiquitinated ERAD substrates to the proteasome, as was suggested (Bays and Hampton 2002). However, this function cannot serve as the one and only obligatory role of p97/Cdc48$^{Ufd1/Npl4}$ also in ERAD, because *CDC48* is dispensable for the proteasomal degradation of N-end rule substrates (Ghislain et al. 1996). Moreover, as discussed below, it appears that p97/Cdc48p plays a specific role in ERAD, which may be executed even prior to extensive polyubiquitination of the substrates (Ye et al. 2003; Elkabetz et al. 2004).

2.3
p97/Cdc48p Provides the Driving Force for Dislocation of Lumenal ERAD Substrates

As discussed above, the proteasome, including its proteolytic activity and Cim5/Rpt1, an AAA-ATPase at the base of the 19S regulatory particle, were reported to play a role in extracting ERAD substrates from the ER (Mayer et al. 1998; Jarosch et al. 2002). However, although CPY* is stabilized in *rpt4R* mutants, the dislocation of this lumenal ERAD substrate appears to be independent of the ATPase activity of Rpt4 (Jarosch et al. 2002). Mechanistically, association of the proteasome with the ER membrane is anticipated to facilitate its function in pulling substrates out of the ER. Indeed, ER-bound proteasomes have been reported in yeast and mammalian cells (Rivett 1998; Enenkel et al. 1998; Hori et al. 1999; Brooks et al. 2000; Hirsch and Ploegh

2000; Elkabetz et al. 2004). No function has been assigned to this subpopulation of the proteasome, yet we report on physical interaction between μs and this specific subpopulation of ER-bound proteasome (Elkabetz et al. 2004) (see Fig. 3). This finding is especially intriguing since the substrate we have used as a bait is the ERAD lumenal substrate μs, which becomes exposed to cytosolic components only when it dislocates or at least emerges from the ER membrane. Nonetheless, our data indicate that the proteolytic activity of the proteasome is dispensable for providing the driving force to the actual passage of μs across the ER membrane (Elkabetz et al. 2004) (see Fig. 3). In the presence of proteasome inhibitors, μs crosses the ER membrane but is not released to the cytosol (Mancini et al. 2000; Elkabetz et al. 2004). Instead, μs accumulates as ubiquitin conjugates at the cytosolic face of the ER, as judged by its complete sensitivity to trypsin digestion (Elkabetz et al. 2004) (see Fig. 3). Hence, by studying the lumenal μs we could dissect the dislocation into two consecutive steps: (1) passage of the substrate across the ER membrane to the ER cytosolic face; (2) release of the substrate from the ER cytosolic face to the cytosol (Elkabetz et al. 2004) (see Fig. 7). Our results indicate that the proteolytic activity of the proteasome is involved only in the release of μs from the cytosolic face of the ER to the cytosol, although we could demonstrate this process only in vitro (Elkabetz et al. 2004) (see Fig. 7, step 6a).

The in vitro reconstruction of the release of μs from isolated microsomes without any added cytosol has indicated that all the cytosolic components essential for ERAD are already bound to the ER membrane (Elkabetz et al. 2004) (see Fig. 7). These include soluble components that are recruited to the ER membrane, because they can be released by salt wash and then replenished by adding back fresh cytosol (Elkabetz et al. 2004). Among these components are the proteasome as well as p97 (Elkabetz et al. 2004) (see Fig. 3). Again, if this AAA-ATPase is playing any role in the dislocation of ERAD substrates, its association with the ER membrane should facilitate this process. Indeed, although the majority of p97 is found free in the cytosol, μs pulls down only the small subpopulation of p97 that is bound to the ER (Elkabetz et al. 2004) (see Fig. 3). The possibility that the p97 that operates in ERAD is ER-bound

Fig. 7 Schematic model. Order of events is marked by *numbered arrows*. Step *5* may represent two consecutive steps, one that is blocked in *CDC48* at 37°C and the other that is blocked in *cdc48-10* at 30°C. For step *6*, the two alternative options represent degradation in association with the ER membrane (*6*) or following release to the cytosol (*6a, 6b*). (From Elkabetz et al. 2004; courtesy of *The Journal of Biological Chemistry*)

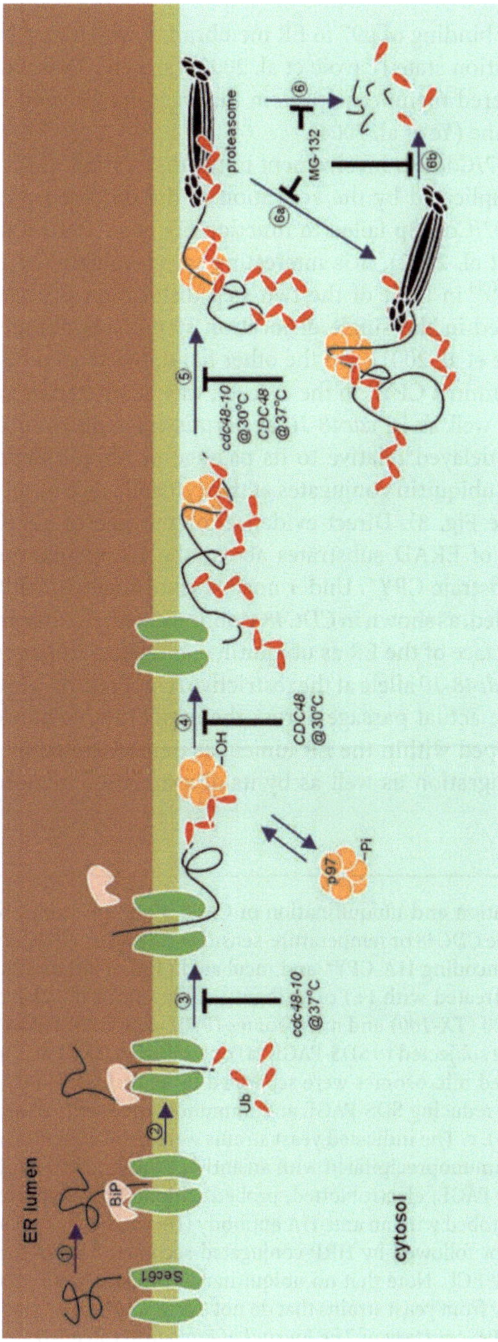
ER lumen
Sec61
BiP
Ub
cytosol
cdc48-10
@37°C
p97
-Pi
-OH
CDC48
@30°C
cdc48-10
@30°C
CDC48
@37°C
proteasome
MG-132

is supported by binding of p97 to ER membranes, which may be regulated by its phosphorylation state (Lavoie et al. 2000) (see Fig. 7) as well as by VIMP, a newly discovered membrane protein that recruits p97 and its partners to the ER membrane (Ye et al. 2004).

Although p97/Cdc48p involvement in the dislocation of ERAD substrates was initially implicated by the activation of the unfolded protein response (UPR) when p97/Cdc48p failed to function (Ye et al. 2001; Rabinovich et al. 2002; Jarosch et al. 2002), it is interesting to re-evaluate the role played by p97/Cdc48$^{\mathrm{Ufd1/Npl4}}$ in light of the two-step dislocation process (see above). p97 is implicated in the single dislocation step of membrane MHC class I heavy chain (Ye et al. 2001). On the other hand, the second dislocation step, the release of luminal CPY* to the cytosol, was hampered in *ufd1-1* (Jarosch et al. 2002), as well as in *cdc48-10* at permissive temperature where CPY* degradation is delayed relative to its passage or ubiquitination, and CPY* accumulates as ubiquitin conjugates at the cytosolic face of the ER (Elkabetz et al. 2004) (see Fig. 8). Direct evidence for the role of p97/Cdc48p in the actual passage of ERAD substrates across the ER membrane is based on the lumenal substrate CPY*. Under normal conditions, the different steps of ERAD are coupled, as shown in *CDC48* strains where CPY* hardly accumulates at the cytosolic face of the ER as ubiquitin conjugates. However, in the strain harboring the *cdc48-10* allele at the restrictive temperature, when Cdc48p fails to function, the actual passage across the membrane is blocked and CPY* remains entrapped within the ER lumen, as demonstrated by its protection from trypsin digestion as well as by its accumulation as nonubiquitinated

Fig. 8a–c Dislocation and ubiquitination of CPY* require Cdc48p. Yeast strains expressing wild type *CDC48* or temperature-sensitive *cdc48-10^{ts}* allele were transformed with a plasmid encoding HA-CPY* and incubated (4 h) at either 30°C or 37°C. Cells were disrupted, treated with (+) or without (–) trypsin in the absence or presence of 1% triton X-100 (*TX-100*) and microsomes (P20) were collected by centrifugation. Microsomes were subjected to SDS-PAGE (**a**) or washed in 0.5 M KCl, and wash supernatant and washed microsomes were separated by centrifugation (**b**). Total proteins were resolved by reducing SDS-PAGE and immunoblotted with an anti-HA antibody (IB: anti-HA; **a**, **b**). **c**. The indicated yeast strains were incubated (4 h) at 30°C or 37°C, HA-CPY* was immunoprecipitated with an anti-HA antibody (IP: anti HA), resolved by reducing SDS-PAGE, electroblotted, probed with an anti-ubiquitin antibody (IB: anti-Ub) and reprobed with an anti-HA antibody (IB: anti-HA). Antibodies were conjugated to HRP or followed by HRP-conjugated secondary antibodies and the HRP was visualized by ECL. Note that no ubiquitinated proteins were precipitated by the anti-HA antibody from yeast strains that do not express HA-CPY* (lanes *9–12*). (From Elkabetz et al. 2004; courtesy of *The Journal of Biological Chemistry*)

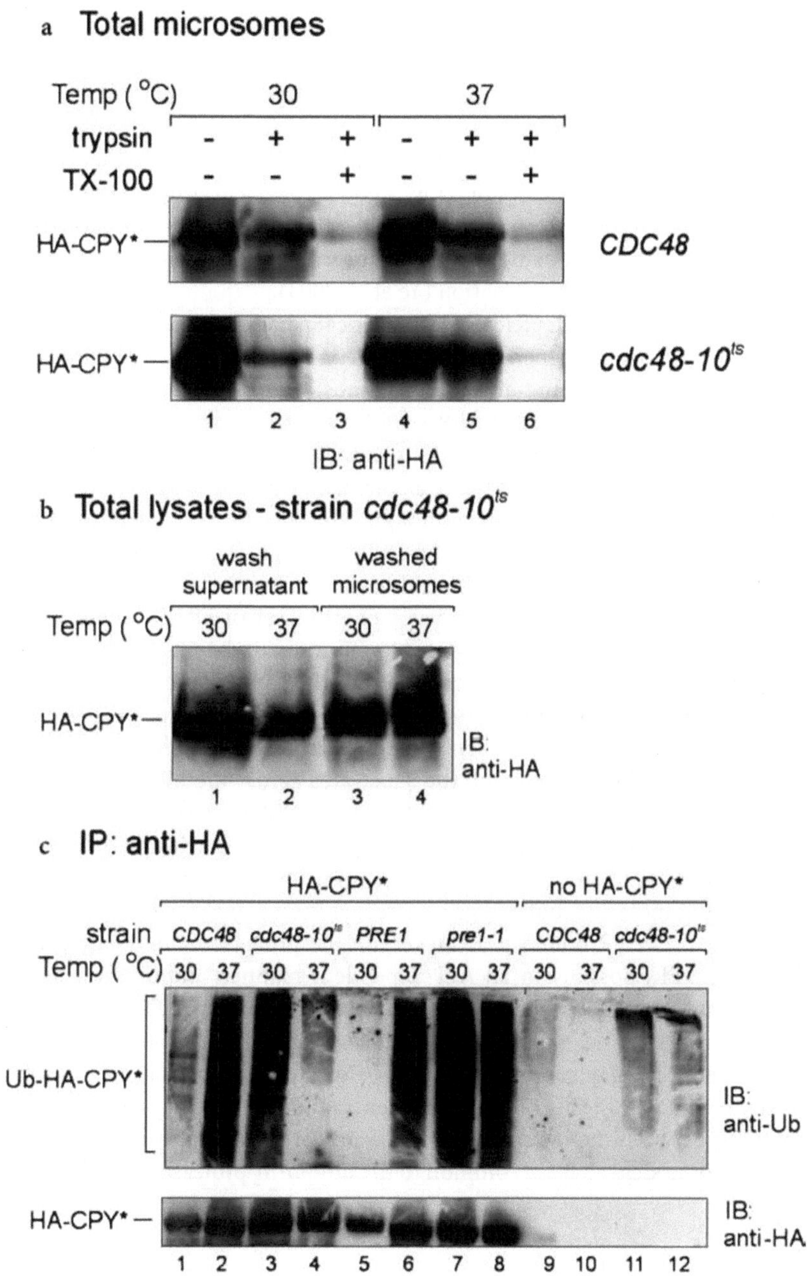
a Total microsomes
Temp (°C) 30 37
trypsin - + + - + +
TX-100 - - + - - +
HA-CPY*
CDC48
HA-CPY*
cdc48-10ts
1 2 3 4 5 6
IB: anti-HA
b Total lysates - strain cdc48-10ts
wash supernatant
washed microsomes
Temp (°C) 30 37 30 37
HA-CPY*
IB: anti-HA
1 2 3 4
c IP: anti-HA
HA-CPY*
no HA-CPY*
strain CDC48 cdc48-10ts PRE1 pre1-1 CDC48 cdc48-10ts
Temp (°C) 30 37 30 37 30 37 30 37 30 37 30 37
Ub-HA-CPY*
IB: anti-Ub
HA-CPY*
IB: anti-HA
1 2 3 4 5 6 7 8 9 10 11 12

protein (Elkabetz et al. 2004) (see Fig. 8). Similarly, although Jarosch et al. suggest that Ufd1p does not operate at the early stages of ERAD but just before the proteasome, their finding that ubiquitinated CPY* does not accumulate in *ufd1-1* cells indicates a delayed dislocation when this partner of Cdc48p is not fully functional (Jarosch et al. 2002). Hence, if p97/Cdc48$^{Ufd1/Npl4}$ complex is involved in the actual passage of CPY* across the ER membrane, before this lumenal substrate has the opportunity to become extensively ubiquitinated, then p97/Cdc48p must recognize substrates that are not yet ubiquitinated. Indeed, a recent report demonstrates dual recognition of substrates by p97, before and after polyubiquitination (Ye et al. 2003).

To conclude, although pleiotropic p97/Cdc48p has additional roles in cells, it is one of the more general components in ERAD, which is required for degradation of membrane and lumenal substrates. Moreover, p97/Cdc48p participated in several steps in ERAD, including the actual passage across the ER membrane, which is unique to ERAD, as well as presenting ubiquitinated substrates to the proteasome, which is shared by other proteasomal substrates (Fig. 7). Finally, although p97/Cdc48p is not dedicated exclusively to ERAD, its ability to physically associate with different sets of partners, ERAD substrates, VIMP, and the E3 gp78 (Zhong et al. 2004) may suggest that this AAA-ATPase acts as a coordinator of ERAD events at the cytosolic face of the ER. The p97/Cdc48p recruited to the ER membrane can bind emerging luminal substrates either in the nonubiquitinated form or following polyubiquitination at least at a single lysine residue (Fig. 7, step 3). Exploiting ATP-dependent conformational changes (Rouiller et al. 2000), ER-bound p97/Cdc48p may pull the substrates across the ER membrane to allow further polyubiquitination to proceed (Fig. 7, step 4). The latter may be assisted by E4/Ufd2, which interacts with p97/Cdc48p (Koegl et al. 1999). Along the lines of this model, polyubiquitin does not appear to serve as a ratcheting molecule, but rather as a recognition signal for p97 (Flierman et al. 2003). The ER-bound p97 can recruit the proteasome to the cytosolic face of the ER, possibly in an ATP-dependent manner, and/or present the polyubiquitinated substrates to the ER-bound proteasome (Fig. 7, steps 5, 6a). The proteasome, which executes the ultimate degradation step in the ERAD pathway, may cooperate with p97/Cdc48p in the release of the substrate to the cytosol (Fig. 7, step 6a). Hence, p97/Cdc48p may maintain the coupling between the different steps in a process as complex as ERAD, including some steps that are common to the ubiquitin-proteasome pathway in general.

References

Alberti S, Bohse K, Arndt V, Schmitz A, Hohfeld J (2004) The Cochaperone HspBP1 inhibits the CHIP ubiquitin ligase and stimulates the maturation of the cystic fibrosis transmembrane conductance regulator. Mol Biol Cell 15:4003–4010

Amitay R, Bar-Nun S, Haimovich J, Rabinovich E, Shachar I (1991) Post-translational regulation of IgM expression in B lymphocytes. J Biol Chem 266:12568–12573

Amitay R, Shachar I, Rabinovich E, Haimovich J, Bar-Nun S (1992) Degradation of secretory IgM in B lymphocytes occurs in a post-endoplasmic reticulum compartment and is mediated by a cysteine protease. J Biol Chem 267:20694–20700

Bays NW, Gardner RG, Seelig LP, Joazeiro CA, Hampton RY (2001a) Hrd1p/Der3p is a membrane-anchored ubiquitin ligase required for ER-associated degradation. Nat Cell Biol 3:24–29

Bays NW, Hampton RY (2002) Cdc48-ufd1-npl4: stuck in the middle with ub. Curr Biol 12:R366–R371

Bays NW, Wilhovsky SK, Goradia A, Hodgkiss-Harlow K, Hampton RY (2001b) HRD4/NPL4 is required for the proteasomal processing of ubiquitinated ER proteins. Mol Biol Cell 12:4114–4128

Biederer T, Volkwein C, Sommer T (1996) Degradation of subunits of the Sec61p complex, an integral component of the ER membrane, by the ubiquitin-proteasome pathway. EMBO J 15:2069–2076

Biederer T, Volkwein C, Sommer T (1997) Role of Cue1p in ubiquitination and degradation at the ER surface [see comments]. Science 278:1806–1809

Bogyo M, McMaster JS, Gaczynska M, Tortorella D, Goldberg AL, Ploegh H (1997) Covalent modification of the active site threonine of proteasomal beta subunits and the Escherichia coli homolog HslV by a new class of inhibitors. Proc Natl Acad Sci U S A 94:6629–6634

Bonifacino JS, Weissman AM (1998) Ubiquitin and the control of protein fate in the secretory and endocytic pathways. Annu Rev Cell Dev Biol 14:19–57

Bordallo J, Plemper RK, Finger A, Wolf DH (1998) Der3p/Hrd1p is required for endoplasmic reticulum-associated degradation of misfolded lumenal and integral membrane proteins. Mol Biol Cell 9:209–222

Braun S, Matuschewski K, Rape M, Thoms S, Jentsch S (2002) Role of the ubiquitin-selective CDC48(UFD1/NPL4) chaperone (segregase) in ERAD of OLE1 and other substrates. EMBO J 21:615–621

Brodsky JL, McCracken AA (1999) ER protein quality control and proteasome-mediated protein degradation. Semin Cell Dev Biol 10:507–513

Brodsky JL, Werner ED, Dubas ME, Goeckeler JL, Kruse KB, McCracken AA (1999) The requirement for molecular chaperones during endoplasmic reticulum-associated protein degradation demonstrates that protein export and import are mechanistically distinct. J Biol Chem 274:3453–3460

Brooks P, Fuertes G, Murray RZ, Bose S, Knecht E, Rechsteiner MC, Hendil KB, Tanaka K, Dyson J, Rivett J (2000) Subcellular localization of proteasomes and their regulatory complexes in mammalian cells. Biochem J 346:155–161

Caldwell SR, Hill KJ, Cooper AA (2001) Degradation of endoplasmic reticulum (ER) quality control substrates requires transport between the ER and Golgi. J Biol Chem 276:23296–23303

Chen C, Bonifacino JS, Yuan YC, Klausner RD (1988) Selective degradation of T-cell antigen receptor chains retained in a pre-Golgi compartment. J Cell Biol 107:2149–2161

Cyr DM, Hohfeld J, Patterson C (2002) Protein quality control: U-box-containing E3 ubiquitin ligases join the fold. Trends Biochem Sci 27:368–375

Dai RM, Chen E, Longo DL, Gorbea CM, Li CC (1998) Involvement of valosin-containing protein, an ATPase co-purified with IkappaBalpha and 26 S proteasome, in ubiquitin-proteasome-mediated degradation of IkappaBalpha. J Biol Chem 273:3562–3573

Dai RM, Li CC (2001) Valosin-containing protein is a multi-ubiquitin chain-targeting factor required in ubiquitin-proteasome degradation. Nat Cell Biol 3:740–744

Elkabetz Y, Kerem A, Tencer L, Winitz D, Kopito RR, Bar-Nun S (2003) Immunoglobulin light chains dictate vesicular transport-dependent and -independent routes for IgM degradation by the ubiquitin-proteasome pathway. J Biol Chem 278:18922–18929

Elkabetz Y, Shapira I, Rabinovich E, Bar-Nun S (2004) Distinct steps in dislocation of luminal endoplasmic reticulum-associated degradation substrates: roles of endoplasmic reticulum-bound p97/Cdc48p and proteasome. J Biol Chem 279:3980–3989

Ellgaard L, Helenius A (2003) Quality control in the endoplasmic reticulum. Nat Rev Mol Cell Biol 4:181–191

Enenkel C, Lehmann A, Kloetzel PM (1998) Subcellular distribution of proteasomes implicates a major location of protein degradation in the nuclear envelope-ER network in yeast. EMBO J 17:6144–6154

Fang S, Ferrone M, Yang C, Jensen JP, Tiwari S, Weissman AM (2001) The tumor autocrine motility factor receptor, gp78, is a ubiquitin protein ligase implicated in degradation from the endoplasmic reticulum. Proc Natl Acad Sci U S A 98:14422–14427

Flierman D, Ye Y, Dai M, Chau V, Rapoport TA (2003) Polyubiquitin serves as a recognition signal, rather than a ratcheting molecule, during retrotranslocation of proteins across the endoplasmic reticulum membrane. J Biol Chem 278:34774–34782

Friedlander R, Jarosch E, Urban J, Volkwein C, Sommer T (2000) A regulatory link between ER-associated protein degradation and the unfolded-protein response. Nat Cell Biol 2:379–384

Galan JM, Cantegrit B, Garnier C, Namy O, Haguenauer-Tsapis R (1998) 'ER degradation' of a mutant yeast plasma membrane protein by the ubiquitin-proteasome pathway. FASEB J 12:315–323

Gardner RG, Hampton RY (1999) A 'distributed degron' allows regulated entry into the ER degradation pathway. EMBO J 18:5994–6004

Ghislain M, Dohmen RJ, Levy F, Varshavsky A (1996) Cdc48p interacts with Ufd3p, a WD repeat protein required for ubiquitin-mediated proteolysis in Saccharomyces cerevisiae. EMBO J 15:4884–4899

Gillece P, Luz JM, Lennarz WJ, de la Cruz FJ, Romisch K (1999) Export of a cysteine-free misfolded secretory protein from the endoplasmic reticulum for degradation requires interaction with protein disulfide isomerase. J Cell Biol 147:1443–1456

Gitan RS, Eide DJ (2000) Zinc-regulated ubiquitin conjugation signals endocytosis of the yeast ZRT1 zinc transporter. Biochem J 346:329–336
Gnann A, Riordan JR, Wolf DH (2004) Cystic fibrosis transmembrane conductance regulator degradation depends on the lectins Htm1p/EDEM and the Cdc48 protein complex in yeast. Mol Biol Cell 15:4125–4135
Gusarova V, Caplan AJ, Brodsky JL, Fisher EA (2001) Apoprotein B degradation is promoted by the molecular chaperones hsp90 and hsp70. J Biol Chem 276:24891–24900
Hammond C, Helenius A (1994) Quality control in the secretory pathway: retention of a misfolded viral membrane glycoprotein involves cycling between the ER, intermediate compartment, and Golgi apparatus. J Cell Biol 126:41–52
Hampton RY, Bhakta H (1997) Ubiquitin-mediated regulation of 3-hydroxy-3-methylglutaryl-CoA reductase. Proc Natl Acad Sci U S A 94:12944–12948
Hampton RY, Gardner RG, Rine J (1996) Role of 26S proteasome and HRD genes in the degradation of 3-hydroxy-3- methylglutaryl-CoA reductase, an integral endoplasmic reticulum membrane protein. Mol Biol Cell 7:2029–2044
Haynes CM, Caldwell S, Cooper AA (2002) An HRD/DER-independent ER quality control mechanism involves Rsp5p- dependent ubiquitination and ER-Golgi transport. J Cell Biol 158:91–101
Hebert DN, Simons JF, Peterson JR, Helenius A (1995) Calnexin, calreticulin, and Bip/Kar2p in protein folding. Cold Spring Harb Symp Quant Biol 60:405–415
Heinemeyer W, Fischer M, Krimmer T, Stachon U, Wolf DH (1997) The active sites of the eukaryotic 20 S proteasome and their involvement in subunit precursor processing. J Biol Chem 272:25200–25209
Heinemeyer W, Kleinschmidt JA, Saidowsky J, Escher C, Wolf DH (1991) Proteinase yscE, the yeast proteasome/multicatalytic-multifunctional proteinase: mutants unravel its function in stress induced proteolysis and uncover its necessity for cell survival. EMBO J 10:555–562
Hendershot LM (1990) Immunoglobulin heavy chain and binding protein complexes are dissociated in vivo by light chain addition. J Cell Biol 111:829–837
Hill K, Cooper AA (2000) Degradation of unassembled Vph1p reveals novel aspects of the yeast ER quality control system. EMBO J 19:550–561
Hiller MM, Finger A, Schweiger M, Wolf DH (1996) ER degradation of a misfolded luminal protein by a cytosolic ubiquitin-proteasome pathway. Science 273:1725–1728
Hirsch C, Ploegh HL (2000) Intracellular targeting of the proteasome. Trends Cell Biol 10:268–272
Hori H, Nembai T, Miyata Y, Hayashi T, Ueno K, Koide T (1999) Isolation and characterization of two 20S proteasomes from the endoplasmic reticulum of rat liver microsomes. J Biochem (Tokyo) 126:722–730
Huyer G, Piluek WF, Fansler Z, Kreft SG, Hochstrasser M, Brodsky JL, Michaelis S (2004) Distinct machinery is required in saccharomyces cerevisiae for the endoplasmic reticulum-associated degradation of a multispanning membrane protein and a soluble luminal protein. J Biol Chem 279:38369–38378
Imai Y, Soda M, Hatakeyama S, Akagi T, Hashikawa T, Nakayama KI, Takahashi R (2002) CHIP is associated with Parkin, a gene responsible for familial Parkinson's disease, and enhances its ubiquitin ligase activity. Mol Cell 10:55–67

Imai Y, Soda M, Inoue H, Hattori N, Mizuno Y, Takahashi R (2001) An unfolded putative transmembrane polypeptide, which can lead to endoplasmic reticulum stress, is a substrate of Parkin. Cell 105:891–902

Jarosch E, Taxis C, Volkwein C, Bordallo J, Finley D, Wolf DH, Sommer T (2002) Protein dislocation from the ER requires polyubiquitination and the AAA-ATPase Cdc48. Nat Cell Biol 4:134–139

Jensen TJ, Loo MA, Pind S, Williams DB, Goldberg AL, Riordan JR (1995) Multiple proteolytic systems, including the proteasome, contribute to CFTR processing. Cell 83:129–135

Kaneko M, Ishiguro M, Niinuma Y, Uesugi M, Nomura Y (2002) Human HRD1 protects against ER stress-induced apoptosis through ER-associated degradation. FEBS Lett 532:147–152

Kikkert M, Doolman R, Dai M, Avner R, Hassink G, van Voorden S, Thanedar S, Roitelman J, Chau V, Wiertz E (2004) Human HRD1 is an E3 ubiquitin ligase involved in degradation of proteins from the endoplasmic reticulum. J Biol Chem 279:3525–3534

Klausner RD, Sitia R (1990) Protein degradation in the endoplasmic reticulum. Cell 62:611–614

Klausner RD, Lippincott-Schwartz J, Bonifacino JS (1990) The T cell antigen receptor: insights into organelle biology. Annu Rev Cell Biol 6:403–431

Knop M, Finger A, Braun T, Hellmuth K, Wolf DH (1996) Der1, a novel protein specifically required for endoplasmic reticulum degradation in yeast. EMBO J 15:753–763

Koegl M, Hoppe T, Schlenker S, Ulrich HD, Mayer TU, Jentsch S (1999) A novel ubiquitination factor, E4, is involved in multiubiquitin chain assembly. Cell 96:635–644

Kopito RR (1997) ER quality control: the cytoplasmic connection. Cell 88:427–430

Lavoie C, Chevet E, Roy L, Tonks NK, Fazel A, Posner BI, Paiement J, Bergeron JJ (2000) Tyrosine phosphorylation of p97 regulates transitional endoplasmic reticulum assembly in vitro. Proc Natl Acad Sci U S A 97:13637–13642

Lee YK, Brewer JW, Hellman R, Hendershot LM (1999) BiP and immunoglobulin light chain cooperate to control the folding of heavy chain and ensure the fidelity of immunoglobulin assembly. Mol Biol Cell 10:2209–2219

Lilley BN, Ploegh HL (2004) A membrane protein required for dislocation of misfolded proteins from the ER. Nature 429:834–840

Loayza D, Tam A, Schmidt WK, Michaelis S (1998) Ste6p mutants defective in exit from the endoplasmic reticulum (ER) reveal aspects of an ER quality control pathway in Saccharomyces cerevisiae. Mol Biol Cell 9:2767–2784

Mancini R, Fagioli C, Fra AM, Maggioni C, Sitia R (2000) Degradation of unassembled soluble Ig subunits by cytosolic proteasomes: evidence that retrotranslocation and degradation are coupled events. FASEB J 14:769–778

Matlack KE, Misselwitz B, Plath K, Rapoport TA (1999) BiP acts as a molecular ratchet during posttranslational transport of prepro-alpha factor across the ER membrane. Cell 97:553–564

Mayer TU, Braun T, Jentsch S (1998) Role of the proteasome in membrane extraction of a short-lived ER transmembrane protein. EMBO J 17:3251–3257

McCracken AA, Brodsky JL (1996) Assembly of ER-associated protein degradation in vitro: dependence on cytosol, calnexin, and ATP. J Cell Biol 132:291–298

Meacham GC, Patterson C, Zhang W, Younger JM, Cyr DM (2001) The Hsc70 co-chaperone CHIP targets immature CFTR for proteasomal degradation. Nat Cell Biol 3:100–105

Medicherla B, Kostova Z, Schaefer A, Wolf DH (2004) A genomic screen identifies Dsk2p and Rad23p as essential components of ER-associated degradation. EMBO Rep 5:692–697

Meerovitch K, Wing S, Goltzman D (1998) Proparathyroid hormone-related protein is associated with the chaperone protein BiP and undergoes proteasome-mediated degradation. J Biol Chem 273:21025–21030

Meusser B, Sommer T (2004) Vpu-mediated degradation of CD4 reconstituted in yeast reveals mechanistic differences to cellular ER-associated protein degradation. Mol Cell 14:247–258

Meyer HH, Wang Y, Warren G (2002) Direct binding of ubiquitin conjugates by the mammalian p97 adaptor complexes, p47 and Ufd1-Npl4. EMBO J 21:5645–5652

Nadav E, Shmueli A, Barr H, Gonen H, Ciechanover A, Reiss Y (2003) A novel mammalian endoplasmic reticulum ubiquitin ligase homologous to the yeast Hrd1. Biochem Biophys Res Commun 303:91–97

Nishikawa S, Fewell SW, Kato Y, Brodsky JL, Endo T (2001) Molecular chaperones in the yeast endoplasmic reticulum maintain the solubility of proteins for retro-translocation and degradation. J Cell Biol 153:1061–1070

Pilon M, Schekman R, Romisch K (1997) Sec61p mediates export of a misfolded secretory protein from the endoplasmic reticulum to the cytosol for degradation. EMBO J 16:4540–4548

Plemper RK, Bohmler S, Bordallo J, Sommer T, Wolf DH (1997) Mutant analysis links the translocon and BiP to retrograde protein transport for ER degradation. Nature 388:891–895

Plemper RK, Egner R, Kuchler K, Wolf DH (1998) Endoplasmic reticulum degradation of a mutated ATP-binding cassette transporter Pdr5 proceeds in a concerted action of Sec61 and the proteasome. J Biol Chem 273:32848–32856

Rabinovich E, Bar-Nun S, Amitay R, Shachar I, Gur B, Taya M, Haimovich J (1993) Different assembly species of IgM are directed to distinct degradation sites along the secretory pathway. J Biol Chem 268:24145–24148

Rabinovich E, Kerem A, Frohlich KU, Diamant N, Bar-Nun S (2002) AAA-ATPase p97/Cdc48p, a cytosolic chaperone required for endoplasmic reticulum-associated protein degradation. Mol Cell Biol 22:626–634

Rape M, Hoppe T, Gorr I, Kalocay M, Richly H, Jentsch S (2001) Mobilization of processed, membrane-tethered SPT23 transcription factor by CDC48(UFD1/NPL4), a ubiquitin-selective chaperone. Cell 107:667–677

Riezman H (1997) The ins and outs of protein translocation. Science 278:1728–1729

Rivett AJ (1998) Intracellular distribution of proteasomes. Curr Opin Immunol 10:110–114

Rock KL, Gramm C, Rothstein L, Clark K, Stein R, Dick L, Hwang D, Goldberg AL (1994) Inhibitors of the proteasome block the degradation of most cell proteins and the generation of peptides presented on MHC class I molecules. Cell 78:761–771

Romisch K (1999) Surfing the Sec61 channel: bidirectional protein translocation across the ER membrane. J Cell Sci 112:4185–4191

Rouiller I, Butel VM, Latterich M, Milligan RA, Wilson-Kubalek EM (2000) A major conformational change in p97 AAA ATPase upon ATP binding. Mol Cell 6:1485–1490

Rubin DM, Glickman MH, Larsen CN, Dhruvakumar S, Finley D (1998) Active site mutants in the six regulatory particle ATPases reveal multiple roles for ATP in the proteasome. EMBO J 17:4909–4919

Shachar I, Amitay R, Rabinovich E, Haimovich J, Bar-Nun S (1992) Polymerization of secretory IgM in B lymphocytes is prevented by a preceding targeting to a degradation pathway. J Biol Chem 267:24241–24247

Sommer T, Jentsch S (1993) A protein translocation defect linked to ubiquitin conjugation at the endoplasmic reticulum. Nature 365:176–179

Spear ED, Ng DT (2003) Stress tolerance of misfolded carboxypeptidase Y requires maintenance of protein trafficking and degradative pathways. Mol Biol Cell 14:2756–2767

Stafford FJ, Bonifacino JS (1991) A permeabilized cell system identifies the endoplasmic reticulum as a site of protein degradation. J Cell Biol 115:1225–1236

Swanson R, Locher M, Hochstrasser M (2001) A conserved ubiquitin ligase of the nuclear envelope/endoplasmic reticulum that functions in both ER-associated and Matalpha2 repressor degradation. Genes Dev 15:2660–2674

Taxis C, Hitt R, Park SH, Deak PM, Kostova Z, Wolf DH (2003) Use of modular substrates demonstrates mechanistic diversity and reveals differences in chaperone requirement of ERAD. J Biol Chem 278:35903–35913

Tiwari S, Weissman AM (2001) Endoplasmic reticulum (ER)-associated degradation of T cell receptor subunits. Involvement of ER-associated ubiquitin-conjugating enzymes (E2s). J Biol Chem 276:16193–16200

Tsai B, Rodighiero C, Lencer WI, Rapoport TA (2001) Protein disulfide isomerase acts as a redox-dependent chaperone to unfold cholera toxin. Cell 104:937–948

Vashist S, Kim W, Belden WJ, Spear ED, Barlowe C, Ng DT (2001) Distinct retrieval and retention mechanisms are required for the quality control of endoplasmic reticulum protein folding. J Cell Biol 155:355–368

Vashist S, Ng DTW (2004) Misfolded proteins are sorted by a sequential checkpoint mechanism of ER quality control. J Cell Biol 165:41–52

Verma R, Chen S, Feldman R, Schieltz D, Yates J, Dohmen J, Deshaies RJ (2000) Proteasomal proteomics: identification of nucleotide-sensitive proteasome-interacting proteins by mass spectrometric analysis of affinity-purified proteasomes. Mol Biol Cell 11:3425–3439

Vogel JP, Misra LM, Rose MD (1990) Loss of BiP/GRP78 function blocks translocation of secretory proteins in yeast. J Cell Biol 110:1885–1895

Wang B, Alam SL, Meyer HH, Payne M, Stemmler TL, Davis DR, Sundquist WI (2003) Structure and ubiquitin interactions of the conserved zinc finger domain of Npl4. J Biol Chem 278:20225–20234

Wang Q, Chang A (2003) Substrate recognition in ER-associated degradation mediated by Eps1, a member of the protein disulfide isomerase family. EMBO J 22:3792–3802

Ward CL, Omura S, Kopito RR (1995) Degradation of CFTR by the ubiquitin-proteasome pathway. Cell 83:121–127

Wiertz EJHJ, Tortorella D, Bogyo M, Yu J, Mothes W, Jones TR, Rapoport TA, Ploegh HL (1996) Sec61-mediated transfer of membrane protein from the endoplasmic reticulum to the proteasome for destruction. Nature 384:432–438

Wilhovsky S, Gardner R, Hampton R (2000) HRD gene dependence of endoplasmic reticulum-associated degradation [In Process Citation]. Mol Biol Cell 11:1697–1708

Wilkinson BM, Tyson JR, Reid PJ, Stirling CJ (2000) Distinct domains within yeast Sec61p involved in post-translational translocation and protein dislocation. J Biol Chem 275:521–529

Winitz D, Shachar I, Elkabetz Y, Amitay R, Samuelov M, Bar-Nun S (1996) Degradation of distinct assembly forms of immunoglobulin M occurs in multiple sites in permeabilized B cells. J Biol Chem 271:27645–27651

Ye Y, Meyer HH, Rapoport TA (2001) The AAA ATPase Cdc48/p97 and its partners transport proteins from the ER into the cytosol. Nature 414:652–656

Ye Y, Meyer HH, Rapoport TA (2003) Function of the p97-Ufd1-Npl4 complex in retro-translocation from the ER to the cytosol: dual recognition of nonubiquitinated polypeptide segments and polyubiquitin chains. J Cell Biol 162:71–84

Ye Y, Shibata Y, Yun C, Ron D, Rapoport TA (2004) A membrane protein complex mediates retro-translocation from the ER lumen into the cytosol. Nature 429:841–847

Yoshida Y, Chiba T, Tokunaga F, Kawasaki H, Iwai K, Suzuki T, Ito Y, Matsuoka K, Yoshida M, Tanaka K, Tai T (2002) E3 ubiquitin ligase that recognizes sugar chains. Nature 418:438–442

Yoshida Y, Tokunaga F, Chiba T, Iwai K, Tanaka K, Tai T (2003) Fbs2 is a new member of the E3 ubiquitin ligase family that recognizes sugar chains. J Biol Chem 278:43877–43884

Zhang Y, Nijbroek G, Sullivan ML, McCracken AA, Watkins SC, Michaelis S, Brodsky JL (2001) Hsp70 molecular chaperone facilitates endoplasmic reticulum-associated protein degradation of cystic fibrosis transmembrane conductance regulator in yeast. Mol Biol Cell 12:1303–1314

Zhong X, Shen Y, Ballar P, Apostolou A, Agami R, Fang S (2004) AAA ATPase p97/VCP interacts with gp78: a ubiquitin ligase for ER-associated degradation. J Biol Chem 279:45676–45684

Zhou M, Schekman R (1999) The engagement of Sec61p in the ER dislocation process. Mol Cell 4:925–934

CTMI (2006) 300:127–148

The Ins and Outs of Intracellular Peptides and Antigen Presentation by MHC Class I Molecules

T. Groothuis · J. Neefjes (✉)

Division of Tumor Biology, The Netherlands Cancer Institute, Plesmanlaan 121, 1066 CX Amsterdam, The Netherlands
j.neefjes@nki.nl

Abstract MHC class I molecules present small intracellular generated fragments to the outside surveying immune system. This is the result of a series of biochemical processes involving biosynthesis, degradation, translocation, intracellular transport, diffusion, and many more. Critical intermediates and end products of this cascade of events are peptides. The peptides are generated by the proteasome, degraded by peptidases unless transported into the ER where another peptidase and MHC class I molecules are waiting. Unless peptides bind to MHC class I molecules, they are released from the ER and enter the cytoplasm by a system resembling the ERAD pathway in many aspects. The cycle of peptides over the ER membrane with the proteasome at the input site and peptidases or MHC class I molecules on the output site are central in the MHC class I antigen presentation pathway and this review.

Abbreviations

ER	Endoplasmic reticulum
MHC	Major histocompatibility complex
TPPII	Tripeptidyl peptidase II
PDI	Protein disulfide isomerase
TOP	Thimet oligopeptidase
LAP	Leucine aminopeptidase
BH	Bleomycin hydrolase
MLC	MHC loading complex
PSA	Puromycin sensitive aminopeptidase
ERAD	ER-associated degradation
CFTR	Cystic fibrosis conductance regulator
MDR	Multidrug resistance
H-chain	Heavy-chain
ERAP	Endoplasmic reticulum aminopeptidase (also ERAAP)
ERAAP	Endoplasmic reticulum aminopeptidase associated with antigen processing (also ERAP)
BiP	Luminal binding protein
TAP	Transporter associated with antigen processing
ABC	ATP binding cassette
GFP	Green fluorescent protein
Hsp	Heat shock protein

1 The Classical Model of Antigen Presentation by MHC Class I Molecules

The immune system uses MHC class I molecules to recognize intracellular residing pathogens. These MHC class I molecules present fragments of cytoplasmic or nuclear proteins of such a pathogen at the plasma membrane. This can only occur after at least three consecutive cell biological processes (Fig. 1). Since fragments of proteins are presented, intracellular degradation

Fig. 1 Antigen presentation by MHC class I molecules. Proteins that are destined for degradation by the proteasome are ubiquitinated. Polyubiquitinated proteins are recognized by proteasomes and cleaved into peptides. The majority of the peptides present in the peptide pool are successively trimmed by amino proteases, of which TPPII is the most dominant one. Peptides are trimmed until they are totally reduced to single amino acids. A small part of the peptides may escape the cytosolic proteases by translocation into the ER by the TAP transporter. Once in the ER lumen, the peptides may bind to MHC class I H-chain/β_2m heterodimers. Peptides bound to these heterodimers form a stable complex that will be transported to the plasma membrane for antigen presentation to surveying cytotoxic T cells (CTLs)

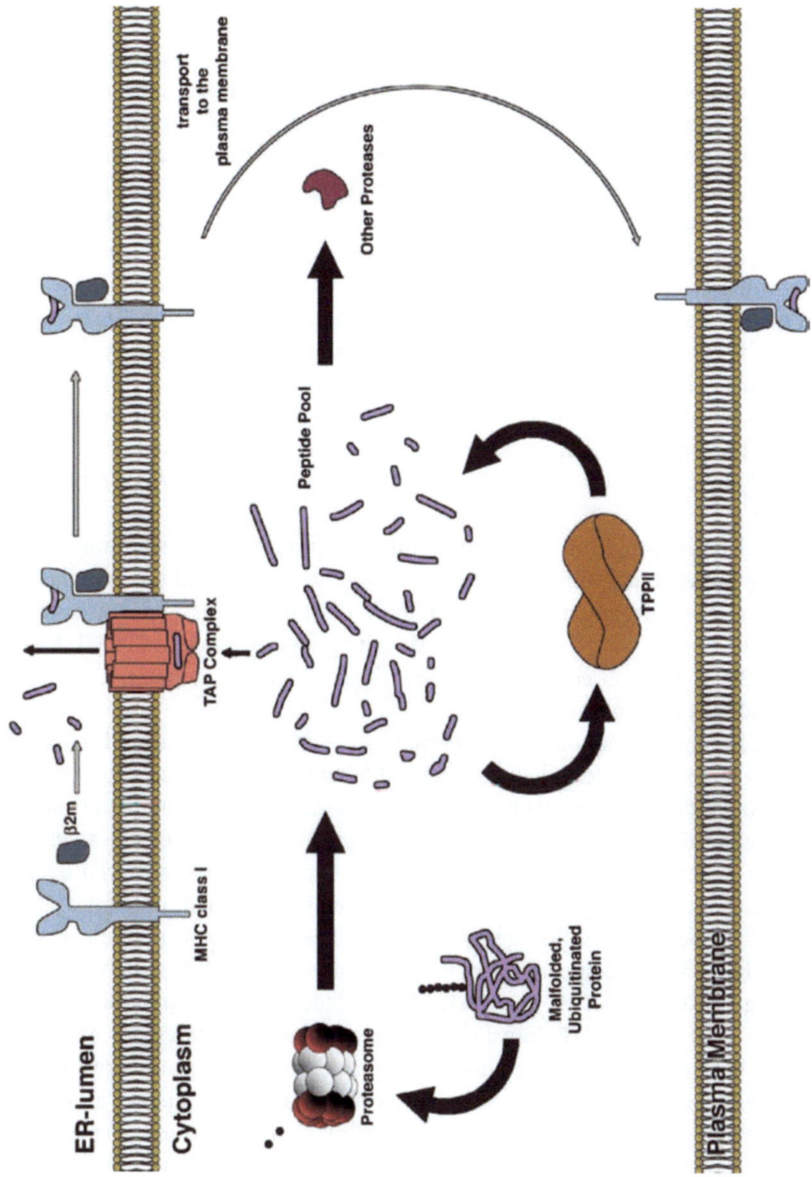
ER-lumen
Cytoplasm
Proteasome
Malfolded,
Ubiquitinated
Protein
MHC class I
β2m
TAP Complex
Peptide Pool
TPPII
Other Proteases
transport
to the
plasma membrane
Plasma Membrane

of proteins to peptide fragments should first occur. The vast majority of intracellular proteins are degraded by the proteasome into small fragments. These proteins may be degraded at the end of their normal life (old proteins), but a large fraction (up to even more than 70%) is degraded immediately after synthesis, probably due to a high failure rate in translation and folding (Reits et al. 2000b; Schubert et al. 2000, reviewed in Yewdell et al. 2003). The latter fraction is termed DriPs and couples viral infection and production of new viral proteins with immediate antigen presentation by MHC class I molecules (Yewdell et al. 2001). The endoplasmic reticulum (ER) of the cell operates a quality control system that identifies misfolded proteins, transports them into the cytoplasm and successively targets them for degradation by the proteasome. Aberrant protein degradation is the mechanism underlying many diseases, including cystic fibrosis and heritable forms of lung and liver disease. The pathways that orchestrate the destruction of aberrant proteins in the ER are collectively termed ER-associated degradation (ERAD) (Hampton 2002).

Still, both the old and the new proteins are degraded into smaller fragments by the proteasome, which is present in both the nucleus and the cytoplasm. Most of these fragments are larger than 15 amino acids and are further degraded by the cytoplasmic peptidase tripeptidylpeptidase II (TPPII, Reits et al. 2004) and subsequently other peptidases (York et al. 2003), until they are reduced to single amino acids.

To become immunologically relevant, the peptides have to bind to MHC class I molecules before they are reduced to single amino acids. The cytosolic protein degradation products are not spontaneously passing membranes, although this is required for association with MHC class I molecules, which are present in the lumen of the ER. Peptide translocation is driven by an ATP-dependent pump located in the ER membrane called TAP, for transporter associated with antigen processing. The pump is a member of the ATP-dependent transporter family that includes drug pumps such as multidrug resistance (MDR) and cystic fibrosis conductance regulator (CFTR) (Klein et al. 1999). TAP translocates many different peptides but excludes those with a modified N-terminus and those containing a proline residue at position 2 or 3 (Momburg et al. 1994b; Neefjes et al. 1993; Neisig et al. 1995). TAP prefers peptides of 8–12 amino acids, but also handles longer peptides albeit less efficiently (Momburg et al. 1994a). Peptides shorter than 8 amino acids are not able to bind to TAP, which makes sense since these are of no immunological relevance because MHC class I molecules require a peptide of minimally 8 amino acids for a stable interaction.

Once in the ER, peptides can have different fates. The immunologically most relevant one is binding to MHC class I H-chain/β_2m heterodimers. These heterodimers are mostly residing in the MHC class I loading complex (MLC), consisting of the peptide transporter, a dedicated chaperone tapasin, and at least two other, more common, chaperones Erp57 and calnexin (Cresswell et al. 1999). Peptide binding to the MHC class I heterodimer releases it from the MLC, allows passage along the ER quality system and transport to the plasma membrane where it represents its cargo (the peptide) to the surveying immune system (Townsend et al. 1989).

However, peptides can also undergo different fates. They can be trimmed by an ER-resident peptidase called ERAP (endoplasmic reticulum aminopeptidase), also named ERAAP (endoplasmic reticulum aminopeptidase associated with antigen processing) (Saric et al. 2002; Serwold et al. 2002; York et al. 2002). This can trim ER resident peptides to the correct size for MHC class I binding and beyond (then destroying MHC class I binding peptides), but probably stops digesting peptides smaller than 8 amino acids.

Peptides can also bind to other ER proteins, mainly the ER chaperones PDI (protein disulfide isomerase), BiP (luminal binding protein), gp96 and gp170 (Lammert et al. 1997b; Spee and Neefjes 1997; Spee et al. 1999). Whether peptides mimic unfolded protein segments and thus bind to chaperones or whether this interaction has another physiological meaning is unclear.

Finally, unbound peptides have to be removed from the ER, which otherwise would obtain high concentrations of peptides that subsequently may affect many other cellular processes. These ER peptides are not released by secretion through the normal secretory route (via ER to Golgi transport) but merely transported back into the cytoplasm by the same machinery as used for ER degradation of proteins employing Sec61/translocon mediated retrotranslocation back to the ER (Koopmann et al. 2000). These different processes ensure successful peptide loading of MHC class I molecules with peptides by employing various old, but highly conserved systems such as protein and peptide degradation, transporters, chaperones, the translocon, and many peptidases again. The end result, peptide presentation by MHC class I molecules, occurs in a very inefficient manner since less than 1% of the proteins degraded deliver a peptide for presentation by MHC class I molecules. The rest is completely turned over and of no immunological significance (Chen et al. 2001; Princiotta et al. 2003; Yewdell et al. 2003).

2
Behavior of Intracellular Peptides

Proteins are degraded by the proteasome in the cytoplasm and in the nucleus. The proteasome diffuses through but not between these compartments since it is excluded from transport through the nuclear pore. TAP is excluded from the nuclear site of the nuclear envelope, which implies that peptides generated in the nucleus have to access the cytoplasm before contacting the peptide transporter TAP (Reits et al. 2003). While diffusing through the various compartments in the cell, many peptides will be trimmed by aminopeptidase activities and only few (less than 2%) will contact TAP and enter the ER (Yewdell et al. 2003).

Peptides will be substrate to peptidases only in a free form. To gain more insight into the longevity of peptides, internally quenched peptides were injected into cells (Reits et al. 2003; Reits et al. 2004). Upon cleavage between the amino acids containing the two groups, the fluorophore is no longer quenched and fluorescence will appear. It turned out that peptides were rapidly and completely degraded within a few seconds. No additional pool of peptide degradation products was detected at later time points. Surprisingly, artificial N-terminal modifications of peptides sufficed to protect them from peptidase activity. This implies that cells contain only aminopeptidases and lack carboxy- and endopeptidase activity. In addition, the proteasome (which is an endopeptidase) does not digest peptides (Reits et al. 2003).

N-terminally protected (and thus stable) fluorescent peptides were introduced in living cells by microinjection. Their rate of diffusion was determined in FRAP experiments because mobility is in approximation proportional to $(\text{mass})^{-1/3}$ (Reits and Neefjes 2001). The protected L-peptides (~1 kDa) moved faster than GFP (green fluorescent proteins, 27 kDa) and GFP moved faster than proteasomes (an intact 20S proteasome is already 700 kDa) (Reits et al. 2003). This implies that the majority of peptide is moving in a free form rather than being associated to other proteins such as heat shock proteins (Hsps). Most likely, these peptides associate transiently to Hsps followed by rapid dissociation.

Do cells have a peptide sink? Closer examination revealed a significant pool of peptides dynamically associated to chromatin (in fact to histones). Whether this influences the MHC class I peptide-loading system in any way is unclear (Reits et al. 2003).

In conclusion, intracellular peptides are mostly free and rapidly moving by normal Brownian motion. In the cytoplasm, TAP and peptidases compete for these peptides. Peptidases are highly active and modify/destroy more than 99% of the TAP substrates, thus strongly reducing the number of peptides entering the ER.

3 Peptides and Peptidases

The proteasome degrades substrate proteins into fragments. In vitro studies suggest that these fragments are peptides of 4–20 amino acids (Cascio et al. 2001). Peptides of 4–7 amino acids are excluded from TAP-driven ER import but longer ones (8–20 amino acids) can be transported into the ER (Koopmann et al. 1996; Momburg et al. 1994a). Peptides have a very short half-life in the cytoplasm/nucleus of intact living cells. Introduction of internally quenched peptides (as mentioned in Sect. 2) into cells to follow their turnover revealed that they are degraded within seconds and exclusively by amino-peptidases (Reits et al. 2003, 2004). The endopeptidase the proteasome is involved in the generation of peptides from proteins, but is irrelevant in peptide degradation. It can be calculated that under normal conditions, the collective cytosolic peptidase activities remove 1.5 amino acid/s (Reits et al. 2004). In other words, a 20-mer peptide will be fully degraded within 15 s but will be irrelevant for the immune system (that is shorter than 8 amino acids) within 8 s. Hence, for a peptide to become immunologically relevant, it should interact with TAP within 8 s after generation by the proteasome, or it will be degraded completely. Many peptides will fail to meet TAP before their destruction, which explains the inefficiency of the MHC class I antigen presentation pathway (Yewdell et al. 2003).

Various cytosolic peptidases have been identified (Fig. 2). These include leucine aminopeptidase (LAP) (Chien et al. 2002; Gu and Walling 2002; Kuo et al. 2003), bleomycin hydrolase (BH) (Nishimura et al. 1989; Nishimura et al. 1987; Sebti et al. 1987; Sebti and Lazo 1987), puromycin sensitive aminopeptidase (PSA) (Hui and Hui 2003; Kakuta et al. 2003; Thompson et al. 2003; Thompson and Hersh LB 2003), thimet oligopeptidase (TOP) (Saric et al. 2004; York et al. 2003), neurolysin (Barelli et al. 1989; Dauch et al. 1991; Millican et al. 1991), and tripeptidylpeptidase II (TPPII) (Balow et al. 1986; Renn et al. 1998). LAP and TOP activity have been shown to affect the peptide pool presented by MHC class I molecules (Saric et al. 2004). TOP as well as neurolysin are probably selective for peptides of 8 up to about 17 amino acids. Whether other peptidases show a defined substrate size selectivity is unclear, with the exception of TPPII (Reits et al. 2004).

TPPII is a huge homomultimeric protease (with a calculated size of 5–9 MDa, which is larger than the proteasome!) (Geier et al. 1999). Reits and colleges have shown that TPPII represents the major proteolytic activity for peptides of 16 amino acids or larger (Reits et al. 2004). Since inhibition of TPPII by a specific compound called butabindide (Breslin et al. 2002) inhibits peptide loading of MHC class I molecules and simultaneous inhibition of

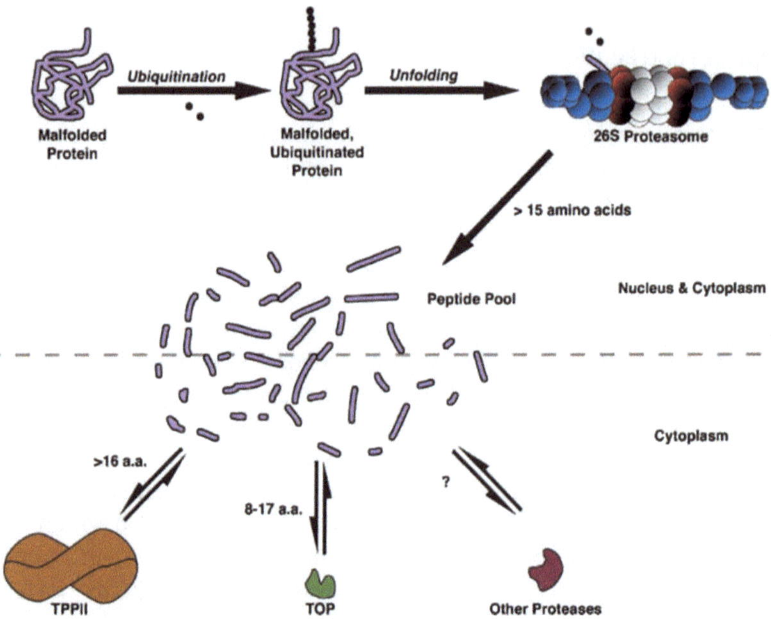

Fig. 2 Antigen processing in the cytoplasm. Before proteins can be recognized by the proteasome, they have to be polyubiquitinated in a process called ubiquitination. After binding of more than four ubiquitins to a single ubiquitin chain the cap of the 26S proteasome (*blue*) is able to bind the ubiquitin chain and the attached substrate protein. The protein is unfolded by the 19S cap and successively degraded by the 20S core into peptides that contain more than 15 amino acids. The peptides form a heterogeneous pool and can diffuse freely through the nucleus and the cytoplasm. In the cytoplasm, the peptides are prone to aminopeptidases with different specificities: TPPII trims peptides that are longer than 16 amino acids; TOP is thought to cleave peptides containing 8–17 amino acids; other proteases, including neurolysin, LAP, and PSA, have a thus far unknown specificity

the proteasome does not further inhibit this, the current concept of peptide generation for MHC class I molecules is in consecutive order:

1. Generation of peptides from proteins by the proteasome. These peptides are mainly larger than 15 amino acids. The proteasome degrades protein substrates into peptide fragments mainly larger than 15 amino acids.
2. TPPII trims these into smaller fragments. This will be achieved by the removal of small (2–3 amino acids long) N-terminal sequences or longer (>8 amino acid) N-terminal fragments. In the latter case, peptides with a new C-terminus are generated for MHC class I molecules.

3. The TPPII substrates will be further trimmed by other peptidases. Possibly TOP is specialized to trim TPPII substrates (it "likes" substrates smaller than ~17 amino acids), but the possible contribution of other peptidases is unclear.

These steps precede peptide import into the ER by a dedicated system, the peptide transporter TAP. A small fraction of peptides (probably less than 1% of peptides and less than 0.01% when starting from protein) survives the collective proteolytic activity by colliding into TAP and are transported to a less hostile environment, the ER lumen. And cytosolic peptidases become relevant again for those peptides failing to bind to MHC class I molecules in the ER lumen, since they will be retrotranslocated back in the cytoplasm for further trimming and destruction.

4 Peptide Import in the Endoplasmic Reticulum: The Transporter Associated with Antigen Processing

Peptides usually do not pass lipid bilayers. Consequently, a dedicated system has been developed to transport cytosolic peptides into the ER for binding to MHC class I molecules (Heemels et al. 1993; Momburg et al. 1994b; Neefjes et al. 1993). A heterodimeric ER-located transporter that performs peptide translocation consists of transporter-associated with antigen processing 1 (TAP1) and TAP2 and is a member of the ATP binding cassette (ABC) transporter superfamily. The TAP1 and TAP2 genes are located in the MHC locus on chromosome 6, very close to two proteasome subunits (Beck et al. 1992; Kelly et al. 1992). The expression of these genes is upregulated by interferon-γ, like the expression levels of MHC H chains and β_2m. Like most ABC transporters, TAP is made of a multimembrane-spanning segment that forms the pore required for actual passage of the ER membrane. This segment is formed by the N-terminal parts of TAP1 and TAP2 and ensures ER retention of the complex. A second area where the peptide is binding then follows this. Both TAP1 and TAP2 contribute to this second segment with TAP2 determining the sequence of the C-terminus, as became apparent by comparing two different rat TAP alleles. Two ATP binding cassettes (ABC), one from TAP1 and the other from TAP2, conclude the transporter. These two ABC domains are essential for the alternating cycles of ATP hydrolysis that drive the different conformational changes in TAP. Probably ATP is first hydrolyzed by TAP1 to open the pore and deliver the peptide in the ER, and then a second ATP is hydrolyzed by TAP2 to close the pore again and return to the ground state (Abele and Tampe 2004; Chen et al. 2004; Reits et al. 2000a; Vos et al. 2000; Vos et al. 1999). These alternating cycles of ATP hydrolysis then drive the conformational changes required for a continuous pumping of peptides. Studies following the lateral mobility of TAP under various conditions have indicated major conformational changes during this cycle. TAP moves "quickly" when inactive and "slowly" when in the process of pumping peptides (Reits et al. 2000b). This may be surprising, but it has been shown that TAP can handle peptides with extended side chains of around 70Å (which is similar to the size of an elongated 8-mer peptide) (Gromme et al. 1997). The mobility assay has been used to monitor the intracellular peptide pool in living cells. This revealed that, under normal conditions, TAP is getting limited amounts of peptides and can handle more substrate. Only upon conditions like stress or virus infection, saturating amounts of peptides are generated. Since most MHC class I molecules are not fully loaded with peptides (and ultimately degraded through the ERAD pathway) (Neefjes and Ploegh 1988), more peptides will enhance the loading and consequently the expression of MHC class I molecules. Probably, a reservoir/excess of MHC class I H-chain/β_2m heterodimers is produced to handle fragments produced under these conditions.

The peptide transporter TAP is unique among the ABC transporter family of pumps. First of all, it is exclusively located in the ER. Second, it acts as a docking protein for a unique chaperone tapasin and MHC class I H-chain/β_2m heterodimers. The heterodimer is further stabilized by other chaperones (ERp57 and calnexin). In fact, probably four tapasin-MHC class I heterodimer-Erp57-calnexin complexes dock onto one TAP protein, thus forming a approximately 1 MDa MHC class I loading complex or MLC (Grandea et al. 1995; Ortmann et al. 1997; Suh et al. 1994). In concepto, this architecture may support efficient peptide loading of MHC class I molecules when peptides imported by TAP are immediately loaded onto the associated MHC class I heterodimers. However, this is likely not too important since:

1. Many peptides require trimming in the ER before being suitable for MHC class I binding (Serwold et al. 2002; York et al. 2002).
2. Many MHC class I alleles are not associated with the TAP complex and are still efficiently loaded (Neisig et al. 1996).
3. MHC class I molecules are still loaded with peptides in tapasin-deficient cells and mice although the quality of the MHC class I associated peptide pool is clearly affected (Brocke et al. 2003; Garbi et al. 2000). However, whether this reflects a difference in chaperoning activity of tapasin or the recruitment to TAP, is unclear.

MHC class I molecules are polymorphic proteins and every human expresses between three and six different alleles. The polymorphic residues cluster around the peptide-binding groove of MHC class I. Consequently, different MHC class I molecules bind different sets of peptides (Elliott et al. 1993; Rammensee et al. 1993). TAP is not polymorphic and thus has to supply all the different MHC class I molecules with peptides. This is only possible when TAP is able to translocate peptides in a fairly sequence-independent manner. Various studies have tested this. TAP translocates peptides with a minimal length of 8 amino acids (MHC class I molecules usually bind peptides of 8 or 9 amino acids). However, peptides of over 40 amino acids are translocated by TAP as well, albeit less efficiently (Momburg et al. 1994a; Schumacher et al. 1994). Although some small differences in sequence selectivity have been observed, TAP translocates peptides with only minor distinction for sequence, with two exceptions. Peptides with the imino acid proline at position 2 or 3 are poorly handled, but still presented by particular MHC class I alleles. In these cases, probably longer peptides (that reposition the proline in the sequence) are translocated by TAP followed by further trimming in the ER lumen (Neisig et al. 1995). Furthermore, selectivity for the C-terminal amino acid residue is found in particular species. Murine TAP as well as a rat TAP allelic form selectively translocate peptides with a hydrophobic or aromatic C-terminal residue, whereas another rat allele and human TAP are very nonselective for amino acids at this position (Momburg et al. 1994b). If the amino acid side chains of a peptide are not recognized by TAP, what is? Further peptide modifications revealed that the N- and C-termini are critical for recognition. Incorporation of amino acid stereoisomers showed that the peptide bond contributed to the interactions with TAP (Gromme et al. 1997). This resembles the situation for peptide binding to MHC class I molecules where most interactions between the MHC class I molecule and peptide are made through hydrogen bonds to the ends of the peptide as well as the peptide's peptide bond (Bouvier and Wiley 1994).

The fact is that TAP is the only peptide transporter in the ER, especially designed to support the MHC class I antigen presentation pathway and in some respects resembling MHC class I. Like other ABC transporters, TAP supports unidirectional transport of its substrate, a peptide.

5 Peptides and Peptidases in the Endoplasmic Reticulum

TAP translocates peptides preferentially of 8–12 amino acids but also longer peptides (Koopmann et al. 1996). Peptides containing a proline at position

2 or 3 are not handled by TAP but these peptides are still found associated to various MHC class I alleles. It is assumed that in these cases N-terminally extended peptides are transported into the ER followed by trimming in the ER by resident peptidases. Peptidase activity has been observed in the ER (Fruci et al. 2001; Roelse et al. 1994) and only recently the corresponding peptidases have been identified and characterized. The ER-aminopeptidase ERAAP (or ERAP1) has been found and is critical in the handling of many peptides in the ER before they can bind to MHC class I molecules (Saric et al. 2002; Serwold et al. 2002; York et al. 2002). In fact, ERAP trims peptidases from the amino terminus until fragments of 8–9 amino acids are left. ERAP thus acts as a sort of molecular ruler, trimming peptides to sizes fitting MHC class I molecules. Interestingly, as mentioned in Sect. 4, peptides containing a proline at position 2 or 3 are not handled by TAP unless the proline residue is repositioned by N-terminal extension. ERAP1 is designed to handle these peptides because it stops further N-terminal trimming when encompassing a proline at position 2 (York et al. 2002). ERAP is a protease and does not "know" which peptides should bind to the resident MHC class I molecules. Consequently it simply trims these peptides and thus creates but also destroys potential MHC class I binding peptides (Saric et al. 2002; York et al. 2002). Whereas ERAP1 is the first ER-located peptidase identified, it is probably not the only one. Other peptidases (called ERAP2, 3 etc.) may also contribute, but their relative contribution still has to be established.

Thus peptides can bind in the ER to MHC class I molecules (if containing the correct anchor residues and length) and ERAP (albeit transiently). Peptides associated with the ER-resident chaperone gp96 have been used for vaccination purposes, even over an MHC barrier (Arnold et al. 1995; Castelli et al. 2004). This suggested that gp96 was able to accumulate the blueprint of peptides before selection by the endogenously expressed MHC class I molecules. To visualize ER proteins able to bind peptides, radioactive labeled peptides with a photo affinity label were introduced in the ER by TAP-mediated import followed by UV-catalyzed cross-linking. Various proteins were found to associate with these peptides, which were identified as the chaperones PDI (protein disulfide isomerase), calnexin, Erp72, gp96, and gp170. Some of these proteins (calnexin, gp170) were subsequently shown to also induce peptide specific immune responses upon vaccination (Lammert et al. 1997a, 1997b; Spee and Neefjes 1997; Spee et al. 1999). That chaperones associate with peptides is not too surprising since peptides can be considered resembling stretches of unfolded protein, the normal substrates for chaperones. Still, PDI was by far most efficient in binding peptides in the ER. The reason for this is unclear. It may be, but this is not shown, that PDI delivers the peptides for consideration by MHC class I molecules. However, PDI may also deliver the peptides to the SEC61/translocon for export out of the ER (see below), especially because PDI has been proposed to be the lid of the SEC61/translocon complex that opens upon nascent protein import (Gillece et al. 1999), but possibly returns peptides for export back into the cytoplasm.

6 Peptide Export from the Endoplasmic Reticulum

Peptides are apparently not degraded to single amino acids in the ER. Consequently, they have to be removed at one point for destruction. Initial experiments following TAP-dependent import showed transient accumulation of peptides unless these obtained N-linked glycans (Koopmann et al. 2000; Shepherd et al. 1993). Further experiments showed that peptides transiently entered the ER microsomal lumen and re-entered the cytoplasm after some 1–3 min, at least in in vitro experiments. Using glycosylation-deficient and normal microsomes, it was shown that peptides were able to enter the ER by TAP transport, were released from the ER through another activity and re-enter the ER again by TAP activity. The limiting factor in this peptide recycling over the ER membrane was the cytosolic peptidase activity that trimmed the peptides to a size too small for TAP handling (less than 8 amino acids) (Roelse et al. 1994). The activity that removed peptides from the ER required ATP, was (unlike TAP) not pH-sensitive and could not be competed with exogenously added peptides (Roelse et al. 1994). Furthermore, a viral inhibitor (ICP47) that inhibited TAP was unable to simultaneously inhibit export as well (Koopmann et al. 2000). Collectively this indicated that an activity different from TAP released the peptides from the ER.

TAP requires triphosphonucleotides, but not necessarily ATP for driving peptide transfer into the ER (Shepherd et al. 1993). Momburg and colleges used this fact to show that peptides can be translocated by TAP in the presence of GTP, but GTP did not drive export (Roelse et al. 1994). In fact, they show that ATP in the ER lumen was required to drive peptide export. Various bacterial toxins enter the ER by retrograde uptake. These toxins use the ERAD system and the SEC61/translocon to enter the cytoplasm where they are toxic. Momburg introduced exotoxin A in the ER lumen of microsomes and showed that this competed with peptide export (Koopmann et al. 2000). Since exotoxin bound the translocon, these data suggested that peptides also used this part of the ERAD system to leave the ER lumen, unless they are captured by the various chaperones.

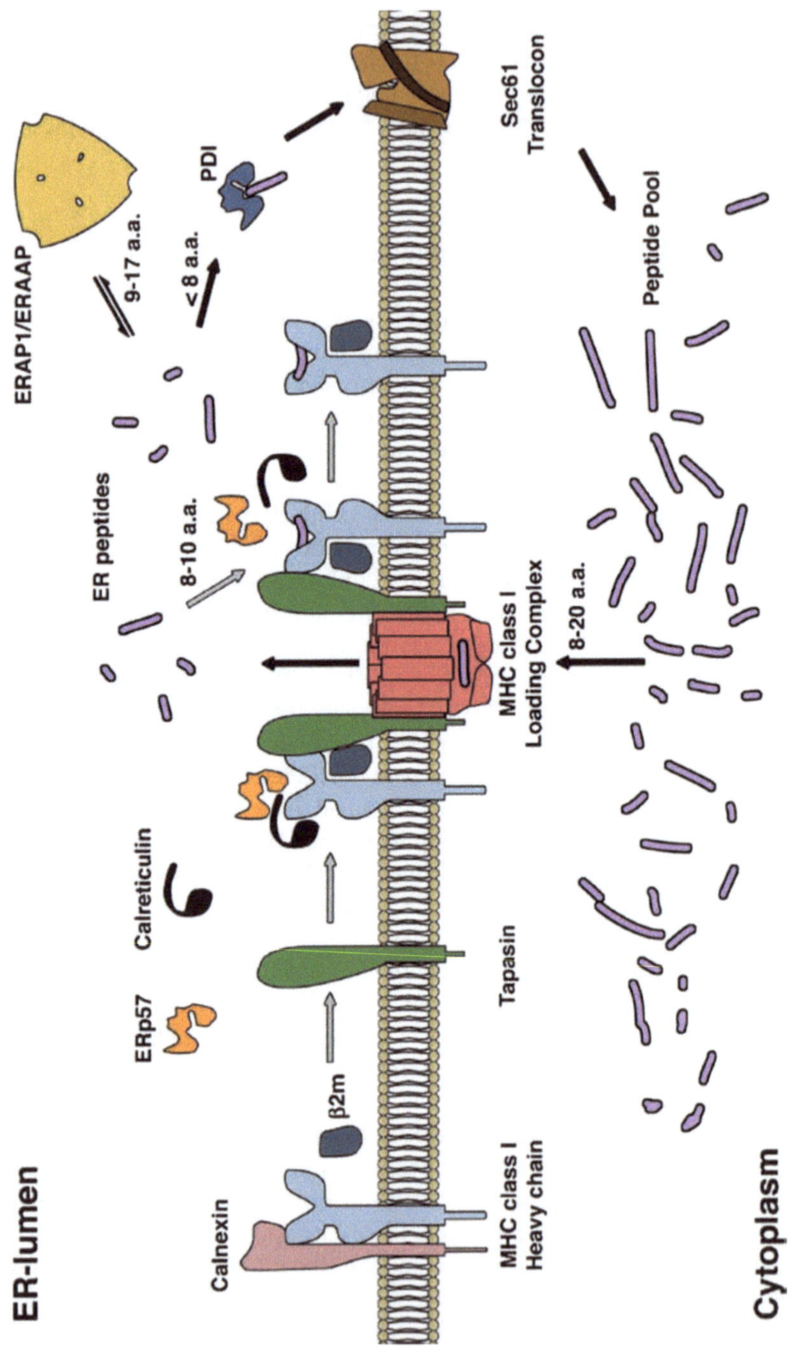
ER-lumen
ERAP1/ERAAP
9-17 a.a.
<8 a.a.
PDI
ER peptides
8-10 a.a.
Calreticulin
ERp57
β2m
Calnexin
MHC class I
Heavy chain
Tapasin
MHC class I
Loading Complex
8-20 a.a.
Sec61
Translocon
Peptide Pool
Cytoplasm

Fig. 3 Peptide cycling over the ER membrane. Cytosolic peptides may escape cytosolic degradation by binding to the peptide transporter TAP. TAP can only bind peptides that are 8–20 amino acids and will transport them in an ATP-dependent process over the ER membrane. Once in the ER, the endoplasmic reticulum aminopeptidase (ERAP), which is specific for peptides of 9–17 amino acids, trims most peptides further. When peptides are not bound to MHC class I molecules, PDI and the Sec61/translocon guide them back to the cytoplasm. Only peptides of 8–10 amino acids with correct anchor residues are able to bind to MHC class I H-chain/β_2m dimers. The heavy chain / β_2m dimer is initially formed with the help of the ER chaperone calnexin, after initial formation the unstable heterodimer is transferred to the MHC class I loading complex (MLC) where it is bound to other ER specific chaperones such as Erp57, calreticulin, and tapasin. Binding of a peptide to the heterodimer stabilizes it and will be granted transport to the plasma membrane by the ER quality control system

The fate of peptides in the ER is:

1. Peptides enter the ER after translocation by TAP. Here peptides have different possibilities (Fig. 3).
2. They bind to MHC class I molecules.
3. They are trimmed by ER aminopeptidases and some of them bind to MHC class I molecules.
4. They have the incorrect sequence for binding to MHC class I molecules and bind ER chaperones.
5. They bind to PDI, which may target them to the SEC61/translocon followed by ATP hydrolysis-driven peptide export back into the cytoplasm.

Peptide cycling over the ER lumen thus resembles the ERAD pathway in many aspects: it requires chaperones, ATP hydrolysis, and retrotranslocation back into the cytoplasm by the SEC61/translocon followed by degradation. Peptides are targeted by cytosolic aminopeptidases; proteins are first degraded by the proteasome and then by cytosolic aminopeptidases.

7 The Equilibrium of Protein and Peptide Degradation, Peptide Cycling and Peptide Binding by MHC Class I Molecules

It seems obvious that the process of protein degradation, peptidase trimming, TAP-mediated peptide translocation, peptide trimming in the ER, peptide loading onto MHC class I molecules, and peptide export are constructed such that optimal peptide generation and loading of MHC class I molecules occurs. This is, however, not the case. The process of antigen presentation, i.e., the resultant of these different steps, is highly inefficient. Recently the kinetics and efficiency of various steps in this process were determined (Princiotta et al. 2003; Yewdell et al. 2003). Proteasomes do not know where to cleave in a protein substrate to generate the correct MHC class I peptides (there are only a standard, nonpolymorphic proteasome and a specialized immunoproteasome, and many different MHC class I molecules). The proteasome probably generates peptides of various lengths but usually the correct C-terminus of a peptide (Cascio et al. 2001). It has to, since cells lack carboxypeptidase activities (Reits et al. 2003). In addition, the proteasome will destroy many potential MHC class I binding peptides.

The resulting proteasomal-produced peptide fragments diffuse through the cell and are targeted first by TPPII and subsequently by other aminopeptidases (Reits et al. 2004). The half-life of many peptides is in the range of 5 s. This activity is such that more than 99% of the peptides generated are destroyed before they can be translocated into the ER (Reits et al. 2003). In the ER, many peptides are destroyed by ERAAP and/or removed by the translocon, while only a few peptides will interact with MHC class I molecules. In fact, cells contain 10^9 proteins of which about 2×10^6 are degraded per minute whereas only 10^4 MHC class I molecules per minute will be loaded with peptides (Yewdell 2001, 2003). This means that only 0.5% or less of the peptides can maximally bind MHC class I molecules. The rest is destroyed unless many more peptides are made than can be loaded onto MHC class I molecules (Reits et al. 2000b). It should be noted that many MHC class I molecules are made but not loaded with peptides. Consequently, under conditions of increased peptide generation (for example following a viral infection), the other MHC class I reservoir can be loaded with peptides as well, resulting in an increased expression of MHC class I molecules. The inefficiency of the system may be essential to generate this pool of peptide recipient MHC class I molecules, lost unless the intracellular peptide pool increases, for example as the result of a viral infection. In conclusion, only few substrates make it into MHC class I binding peptides. The rest are fully turned over into amino acids, not being of any immunological relevance.

8
How the MHC Class I Route Is Manipulated

Obviously, it is fairly beneficial for pathogens to interfere with the system of antigen presentation by MHC class I molecules. This indeed often happens, as discussed in detail elsewhere in this volume. The major viral targets are proteins that are dispensable for household processes. More specifically, viral TAP inhibitors are frequently found, inhibitors affecting the dedicated chaperone tapasin and viral inhibitors targeting MHC class I H-chains or H-chain/β_2m heterodimers for degradation, usually through the ERAD system (Basta et al. 2002; Ben-Arieh et al. 2001; Hewitt et al. 2001; Koppers-Lalic et al. 2003). Viral inhibitors affecting the proteasome, TPPII, TOP or other peptidases, ERAP, ER chaperones, or the SEC61/translocon have not been defined, not unexpectedly, because inhibition of these would affect cell viability, thus being a disadvantage for the pathogens rather than an advantage.

Obviously, chemical inhibitors may be used to affect antigen presentation by MHC class I molecules. Inhibitors for TAP have been designed (being peptides with long side chains), but these cannot be used in living cells or organisms because they do not pass membranes (Gromme et al. 1997). This is also true for the active 35-amino acid fragment of the viral TAP inhibitor ICP47 (Galocha et al. 1997). Inhibitors for the proteasome have been designed and clinically used in anti-cancer treatments (Kane et al. 2003; Kondagunta et al. 2004; Papandreou and Logothetis 2004). These inhibitors prevent proteasome-mediated protein degradation (possibly generating aggregates as a consequence) and peptide loading of MHC class I molecules. Whether these compounds can be used to inhibit/treat MHC class I-related autoimmune diseases such as Bechterew and Reiter's syndrome, is unclear.

9
The Eternal Cycle of Events, or Not ...

In conclusion, proteins are degraded by the proteasome and further degraded by TPPII and other peptidases into single amino acids. These single amino acids can then be used to build new proteins that—immediately in the case of DriPs or slowly for successful proteins—are degraded again into single amino acids. The MHC class I antigen presentation system samples a small amount out of this cycle for presentation purposes. Degradation intermediates that are simultaneously substrate to peptidases in the cytoplasm and ER lumen and to MHC class I molecules are presented. Critical steps in acquiring peptides by MHC class I molecules are the peptide cycle over the ER membrane with TAP pumping peptides into the ER and the SEC61/translocon transporting them out. Although the dynamics of this cycle are not fully understood, the end result is again that a small fraction of imported peptides make it into something immunologically useful: an MHC class I peptide complex. The remainder is ultimately degraded into single amino acids, as are MHC class I molecules failing to obtain peptides during their biosynthesis.

References

Abele R, Tampe R (2004) The ABCs of immunology: structure and function of TAP, the transporter associated with antigen processing. Physiology (Bethesda) 19:216–224

Arnold D, Faath S, Rammensee H, Schild H (1995) Cross-priming of minor histocompatibility antigen-specific cytotoxic T cells upon immunization with the heat shock protein gp96. J Exp Med 182:885–889

Balow RM, Tomkinson B, Ragnarsson U, Zetterqvist O (1986) Purification, substrate specificity, classification of tripeptidyl peptidase II. J Biol Chem 261:2409–2417

Barelli H, Ahmad S, Kostka P, Fox JA, Daniel EE, Vincent JP, Checler F (1989) Neuropeptide-hydrolysing activities in synaptosomal fractions from dog ileum myenteric, deep muscular and submucous plexi. Their participation in neurotensin inactivation. Peptides 10:1055–1061

Basta S, Chen W, Bennink JR, Yewdell JW (2002) Inhibitory effects of cytomegalovirus proteins US2 and US11 point to contributions from direct priming and cross-priming in induction of vaccinia virus-specific CD8(+) T cells. J Immunol 168:5403–5408

Beck S, Kelly A, Radley E, Khurshid F, Alderton RP, Trowsdale J (1992) DNA sequence analysis of 66 kb of the human MHC class II region encoding a cluster of genes for antigen processing. J Mol Biol 228:433–441

Ben-Arieh SV, Zimerman B, Smorodinsky NI, Yaacubovicz M, Schechter C, Bacik I, Gibbs J, Bennink JR, Yewdell JW, Coligan JE et al (2001) Human cytomegalovirus protein US2 interferes with the expression of human HFE, a nonclassical class I major histocompatibility complex molecule that regulates iron homeostasis. J Virol 75:10557–10562

Bouvier M, Wiley DC (1994) Importance of peptide amino and carboxyl termini to the stability of MHC class I molecules. Science 265:398–402

Breslin HJ, Miskowski TA, Kukla MJ, Leister WH, De Winter HL, Gauthier DA, Somers MV, Peeters DC, Roevens PW (2002) Design, synthesis, tripeptidyl peptidase II inhibitory activity of a novel series of (S)-2,3-dihydro-2-(4-alkyl-1H-imidazol-2-yl)-1H-indoles. J Med Chem 45:5303–5310

Brocke P, Armandola E, Garbi N, Hammerling GJ (2003) Downmodulation of antigen presentation by H2-O in B cell lines and primary B lymphocytes. Eur J Immunol 33:411–421

Cascio P, Hilton C, Kisselev AF, Rock KL, Goldberg AL (2001) 26S proteasomes and immunoproteasomes produce mainly N-extended versions of an antigenic peptide. EMBO J 20:2357–2366

Castelli C, Rivoltini L, Rini F, Belli F, Testori A, Maio M, Mazzaferro V, Coppa J, Srivastava PK, Parmiani G (2004) Heat shock proteins: biological functions and clinical application as personalized vaccines for human cancer. Cancer Immunol Immunother 53:227–233

Chen M, Abele R, Tampe R (2004) Functional non-equivalence of ABC signature motifs in the transporter associated with antigen processing (TAP). J Biol Chem 279:46073–46081

Chen W, Bennink JR, Yewdell JW (2001) Quantitating presentation of MHC class I-restricted antigens. Methods Mol Biol 156:245–254

Chien HC, Lin LL, Chao SH, Chen CC, Wang WC, Shaw CY, Tsai YC, Hu HY, Hsu WH (2002) Purification, characterization, genetic analysis of a leucine aminopeptidase from Aspergillus sojae. Biochim Biophys Acta 1576:119–126

Cresswell P, Bangia N, Dick T, Diedrich G (1999) The nature of the MHC class I peptide loading complex. Immunol Rev 172:21–28

Dauch P, Barelli H, Vincent JP, Checler F (1991) Fluorimetric assay of the neurotensin-degrading metalloendopeptidase, endopeptidase 24.16. Biochem J 280:421–426

Elliott T, Smith M, Driscoll P, McMichael A (1993) Peptide selection by class I molecules of the major histocompatibility complex. Curr Biol 3:854–866

Fruci D, Niedermann G, Butler RH, van Endert PM (2001) Efficient MHC class I-independent amino-terminal trimming of epitope precursor peptides in the endoplasmic reticulum. Immunity 15:467–476

Galocha B, Hill A, Barnett BC, Dolan A, Raimondi A, Cook RF, Brunner J, McGeoch DJ, Ploegh HL (1997) The active site of ICP47, a herpes simplex virus-encoded inhibitor of the major histocompatibility complex (MHC)-encoded peptide transporter associated with antigen processing (TAP), maps to the NH2-terminal 35 residues. J Exp Med 185:1565–1572

Garbi N, Tan P, Diehl AD, Chambers BJ, Ljunggren HG, Momburg F, Hammerling GJ (2000) Impaired immune responses and altered peptide repertoire in tapasin-deficient mice. Nat Immunol 1:234–238

Geier E, Pfeifer G, Wilm M, Lucchiari-Hartz M, Baumeister W, Eichmann K, Niedermann G (1999) A giant protease with potential to substitute for some functions of the proteasome. Science 283:978–981

Gillece P, Luz JM, Lennarz WJ, de La Cruz FJ, Romisch K (1999) Export of a cysteine-free misfolded secretory protein from the endoplasmic reticulum for degradation requires interaction with protein disulfide isomerase. J Cell Biol 147:1443–1456

Grandea AG 3rd, Androlewicz MJ, Athwal RS, Geraghty DE, Spies T (1995) Dependence of peptide binding by MHC class I molecules on their interaction with TAP. Science 270:105–108

Gromme M, van der Valk R, Sliedregt K, Vernie L, Liskamp R, Hammerling G, Koopmann JO, Momburg F, Neefjes J (1997) The rational design of TAP inhibitors using peptide substrate modifications and peptidomimetics. Eur J Immunol 27:898–904

Gu YQ, Walling LL (2002) Identification of residues critical for activity of the wound-induced leucine aminopeptidase (LAP-A) of tomato. Eur J Biochem 269:1630–1640

Hampton RY (2002) ER-associated degradation in protein quality control and cellular regulation. Curr Opin Cell Biol 14:476–482

Heemels MT, Schumacher TN, Wonigeit K, Ploegh HL (1993) Peptide translocation by variants of the transporter associated with antigen processing. Science 262:2059–2063

Hewitt EW, Gupta SS, Lehner PJ (2001) The human cytomegalovirus gene product US6 inhibits ATP binding by TAP. EMBO J 20:387–396

Hui M, Hui KS (2003) Neuron-specific aminopeptidase and puromycin-sensitive aminopeptidase in rat brain development. Neurochem Res 28:855–860

Kakuta H, Koiso Y, Nagasawa K, Hashimoto Y (2003) Fluorescent bioprobes for visualization of puromycin-sensitive aminopeptidase in living cells. Bioorg Med Chem Lett 13:83–86

Kane RC, Bross PF, Farrell AT, Pazdur R (2003) Velcade: U.S. FDA approval for the treatment of multiple myeloma progressing on prior therapy. Oncologist 8:508–513

Kelly A, Powis SH, Kerr LA, Mockridge I, Elliott T, Bastin J, Uchanska-Ziegler B, Ziegler A, Trowsdale J, Townsend A (1992) Assembly and function of the two ABC transporter proteins encoded in the human major histocompatibility complex. Nature 355:641–644

Klein I, Sarkadi B, Varadi A (1999) An inventory of the human ABC proteins. Biochim Biophys Acta 1461:237–262

Kondagunta GV, Drucker B, Schwartz L, Bacik J, Marion S, Russo P, Mazumdar M, Motzer RJ (2004) Phase II trial of bortezomib for patients with advanced renal cell carcinoma. J Clin Oncol 22:3720–3725

Koopmann JO, Albring J, Huter E, Bulbuc N, Spee P, Neefjes J, Hammerling GJ, Momburg F (2000) Export of antigenic peptides from the endoplasmic reticulum intersects with retrograde protein translocation through the Sec61p channel. Immunity 13:117–127

Koopmann JO, Post M, Neefjes JJ, Hammerling GJ, Momburg F (1996) Translocation of long peptides by transporters associated with antigen processing (TAP). Eur J Immunol 26:1720–1728

Koppers-Lalic D, Rychlowski M, van Leeuwen D, Rijsewijk FA, Ressing ME, Neefjes JJ, Bienkowska-Szewczyk K, Wiertz EJ (2003) Bovine herpesvirus 1 interferes with TAP-dependent peptide transport and intracellular trafficking of MHC class I molecules in human cells. Arch Virol 148:2023–2037

Kuo LY, Hwang GY, Lai YJ, Yang SL, Lin LL (2003) Overexpression, purification, characterization of the recombinant leucine aminopeptidase II of Bacillus stearothermophilus. Curr Microbiol 47:40–45

Lammert E, Arnold D, Nijenhuis M, Momburg F, Hammerling GJ, Brunner J, Stevanovic S, Rammensee HG, Schild H (1997a) The endoplasmic reticulum-resident stress protein gp96 binds peptides translocated by TAP. Eur J Immunol 27:923–927

Lammert E, Stevanovic S, Brunner J, Rammensee HG, Schild H (1997b) Protein disulfide isomerase is the dominant acceptor for peptides translocated into the endoplasmic reticulum. Eur J Immunol 27:1685–1690

Millican PE, Kenny AJ, Turner AJ (1991) Purification and properties of a neurotensin-degrading endopeptidase from pig brain. Biochem J 276:583–591

Momburg F, Roelse J, Hammerling GJ, Neefjes JJ (1994a) Peptide size selection by the major histocompatibility complex-encoded peptide transporter. J Exp Med 179:1613–1623

Momburg F, Roelse J, Howard JC, Butcher GW, Hammerling GJ, Neefjes JJ (1994b) Selectivity of MHC-encoded peptide transporters from human, mouse and rat. Nature 367:648–651

Neefjes JJ, Momburg F, Hammerling GJ (1993) Selective and ATP-dependent translocation of peptides by the MHC-encoded transporter. Science 261:769–771

Neefjes JJ, Ploegh HL (1988) Allele and locus-specific differences in cell surface expression and the association of HLA class I heavy chain with beta 2-microglobulin: differential effects of inhibition of glycosylation on class I subunit association. Eur J Immunol 18:801–810

Neisig A, Roelse J, Sijts AJ, Ossendorp F, Feltkamp MC, Kast WM, Melief CJ, Neefjes JJ (1995) Major differences in transporter associated with antigen presentation (TAP)-dependent translocation of MHC class I-presentable peptides and the effect of flanking sequences. J Immunol 154:1273–1279

Neisig A, Wubbolts R, Zang X, Melief C, Neefjes J (1996) Allele-specific differences in the interaction of MHC class I molecules with transporters associated with antigen processing. J Immunol 156:3196–3206
Nishimura C, Suzuki H, Tanaka N, Yamaguchi H (1989) Bleomycin hydrolase is a unique thiol aminopeptidase. Biochem Biophys Res Commun 163:788–796
Nishimura C, Tanaka N, Suzuki H (1987) Purification of bleomycin hydrolase with a monoclonal antibody and its characterization. Biochemistry 26:1574–1578
Ortmann B, Copeman J, Lehner PJ, Sadasivan B, Herberg JA, Grandea AG, Riddell SR, Tampe R, Spies T, Trowsdale J, Cresswell P (1997) A critical role for tapasin in the assembly and function of multimeric MHC class I-TAP complexes. Science 277:1306–1309
Papandreou CN, Logothetis CJ (2004) Bortezomib as a potential treatment for prostate cancer. Cancer Res 64:5036–5043
Princiotta MF, Finzi D, Qian SB, Gibbs J, Schuchmann S, Buttgereit F, Bennink JR, Yewdell JW (2003) Quantitating protein synthesis, degradation, endogenous antigen processing. Immunity 18:343–354
Rammensee HG, Falk K, Rotzschke O (1993) Peptides naturally presented by MHC class I molecules. Annu Rev Immunol 11:213–244
Reits E, Griekspoor A, Neijssen J, Groothuis T, Jalink K, van Veelen P, Janssen H, Calafat J, Drijfhout JW, Neefjes J (2003) Peptide diffusion, protection, degradation in nuclear and cytoplasmic compartments before antigen presentation by MHC class I. Immunity 18:97–108
Reits E, Neijssen J, Herberts C, Benckhuijsen W, Janssen L, Drijfhout JW, Neefjes J (2004) A major role for TPPII in trimming proteasomal degradation products for MHC class I antigen presentation. Immunity 20:495–506
Reits EA, Griekspoor AC, Neefjes J (2000a) How does TAP pump peptides? Insights from DNA repair and traffic ATPases. Immunol Today 21:598–600
Reits EA, Neefjes JJ (2001) From fixed to FRAP: measuring protein mobility and activity in living cells. Nat Cell Biol 3:E145–E147
Reits EA, Vos JC, Gromme M, Neefjes J (2000b) The major substrates for TAP in vivo are derived from newly synthesized proteins. Nature 404:774–778
Renn SC, Tomkinson B, Taghert PH (1998) Characterization and cloning of tripeptidyl peptidase II from the fruit fly, Drosophila melanogaster. J Biol Chem 273:19173–19182
Roelse J, Gromme M, Momburg F, Hammerling G, Neefjes J (1994) Trimming of TAP-translocated peptides in the endoplasmic reticulum and in the cytosol during recycling. J Exp Med 180:1591–1597
Saric T, Chang SC, Hattori A, York IA, Markant S, Rock KL, Tsujimoto M, Goldberg AL (2002) An IFN-gamma-induced aminopeptidase in the ER, ERAP1, trims precursors to MHC class I-presented peptides. Nat Immunol 3:1169–1176
Saric T, Graef CI, Goldberg AL (2004) Pathway for degradation of peptides generated by proteasomes: a key role for thimet oligopeptidase and other metallopeptidases. J Biol Chem 279:46273–46732
Schubert U, Anton LC, Gibbs J, Norbury CC, Yewdell JW, Bennink JR (2000) Rapid degradation of a large fraction of newly synthesized proteins by proteasomes. Nature 404:770–774

Schumacher TN, Kantesaria DV, Heemels MT, Ashton-Rickardt PG, Shepherd JC, Fruh K, Yang Y, Peterson PA, Tonegawa S, Ploegh HL (1994) Peptide length and sequence specificity of the mouse TAP1/TAP2 translocator. J Exp Med 179:533–540
Sebti SM, DeLeon JC, Lazo JS (1987) Purification, characterization, amino acid composition of rabbit pulmonary bleomycin hydrolase. Biochemistry 26:4213–4219
Sebti SM, Lazo JS (1987) Separation of the protective enzyme bleomycin hydrolase from rabbit pulmonary aminopeptidases. Biochemistry 26:432–437
Serwold T, Gonzalez F, Kim J, Jacob R, Shastri N (2002) ERAAP customizes peptides for MHC class I molecules in the endoplasmic reticulum. Nature 419:480–483
Shepherd JC, Schumacher TN, Ashton-Rickardt PG, Imaeda S, Ploegh HL, Janeway CA, Jr, Tonegawa S (1993) TAP1-dependent peptide translocation in vitro is ATP dependent and peptide selective. Cell 74:577–584
Spee P, Neefjes J (1997) TAP-translocated peptides specifically bind proteins in the endoplasmic reticulum, including gp96, protein disulfide isomerase and calreticulin. Eur J Immunol 27:2441–2449
Spee P, Subjeck J, Neefjes J (1999) Identification of novel peptide binding proteins in the endoplasmic reticulum: ERp72, calnexin, grp170. Biochemistry 38:10559–10566
Suh WK, Cohen-Doyle MF, Fruh K, Wang K, Peterson PA, Williams DB (1994) Interaction of MHC class I molecules with the transporter associated with antigen processing. Science 264:1322–1326
Thompson MW, Govindaswami M, Hersh LB (2003) Mutation of active site residues of the puromycin-sensitive aminopeptidase: conversion of the enzyme into a catalytically inactive binding protein. Arch Biochem Biophys 413:236–242
Thompson MW, Hersh LB (2003) Analysis of conserved residues of the human puromycin-sensitive aminopeptidase. Peptides 24:1359–1365
Townsend A, Ohlen C, Bastin J, Ljunggren HG, Foster L, Karre K (1989) Association of class I major histocompatibility heavy and light chains induced by viral peptides. Nature 340:443–448
Vos JC, Reits EA, Wojcik-Jacobs E, Neefjes J (2000) Head-head/tail-tail relative orientation of the pore-forming domains of the heterodimeric ABC transporter TAP. Curr Biol 10:1–7
Vos JC, Spee P, Momburg F, Neefjes J (1999) Membrane topology and dimerization of the two subunits of the transporter associated with antigen processing reveal a three-domain structure. J Immunol 163:6679–6685
Yewdell JW (2001) Not such a dismal science: the economics of protein synthesis, folding, degradation and antigen processing. Trends Cell Biol 11:294–297
Yewdell JW, Reits E, Neefjes J (2003) Making sense of mass destruction: quantitating MHC class I antigen presentation. Nat Rev Immunol 3:952–961
Yewdell JW, Schubert U, Bennink JR (2001) At the crossroads of cell biology and immunology: DRiPs and other sources of peptide ligands for MHC class I molecules. J Cell Sci 114:845–851
York IA, Chang SC, Saric T, Keys JA, Favreau JM, Goldberg AL, Rock KL (2002) The ER aminopeptidase ERAP1 enhances or limits antigen presentation by trimming epitopes to 8–9 residues. Nat Immunol 3:1177–1184
York IA, Mo AX, Lemerise K, Zeng W, Shen Y, Abraham CR, Saric T, Goldberg AL, Rock KL (2003) The cytosolic endopeptidase, thimet oligopeptidase, destroys antigenic peptides and limits the extent of MHC class I antigen presentation. Immunity 18:429–440

CTMI (2006) 300:149–168

Entry of Protein Toxins into Mammalian Cells by Crossing the Endoplasmic Reticulum Membrane: Co-opting Basic Mechanisms of Endoplasmic Reticulum-Associated Degradation

J. M. Lord[1] (✉) · L. M. Roberts[1] · W. I. Lencer[2]

[1]Department of Biological Sciences, University of Warwick, Coventry CV4 7AL, UK
Mike.Lord@warwick.ac.uk

[2]The Division of Gastroenterology, Childrens Hospital Boston and the Harvard Digestive Diseases Center and the Department of Pediatrics, Harvard Medical School, Boston, MA 02115, USA

Abstract The catalytic polypeptides of certain bacterial and plant protein toxins reach their substrates in the cytosol of mammalian cells by retro-translocation from the endoplasmic reticulum (ER). Emerging evidence indicates that these proteins subvert the ER-associated protein degradation (ERAD) pathway that normally removes misfolded or unassembled proteins from the ER, to achieve retrotranslocation. Upon entering the ER lumen, the toxins are unfolded to be perceived as ERAD substrates. Toxins that retro-translocate from the ER have an unusually low lysine content to avoid ubiquitin-mediated proteasomal degradation. This allows the exported toxins to refold into the proteasome-resistant, biologically active conformation, and leads to cellular intoxication.

Abbreviations

CT	Cholera toxin
CT-A	Cholera toxin A chain
Stx	SHIGA toxin
ExoA	*Pseudomonas aeruginosa* exotoxin A
RTA	Ricin toxin A chain
RTB	Ricin toxin B chain
Gb3	Globotriaosylceramide
EE	Early endosome
RE	Recycling endosomes
TGN	*trans*-Golgi network
ER	Endoplasmic reticulum
PDI	Protein disulfide isomerase

1 Introduction

Certain bacteria and plants produce protein toxins that introduce a portion of the toxin into the cytosol of target cells to disrupt essential cellular processes such as signalling, cytoskeleton assembly, vesicular trafficking or protein synthesis. Because these toxins act catalytically, they must cross a cellular membrane to interact with their cytosolic substrates. This is not an easy task for any stably folded protein, and the toxins have evolved several mechanisms to breech the cellular barriers.

One way is for a toxin to enter the cell by crossing the plasma membrane. Examples of this type of cell entry would be *Pertussis* adenylate cyclase toxin that causes pertussis (Sekura et al. 1985), *Staphylococcus* A toxin that causes scarlet fever (Bhakdi and Tranum-Jensen 1991), and aerolysin O toxin that induces diarrhea (Buckley et al. 1981). Exactly how these proteins form pores or cross the plasma membrane is poorly understood. Another way is for the toxins to cross the endosomal membrane. Examples of this type of cell entry are diphtheria toxin (Pappenheimer 1977), anthrax toxin (Mock and Fouet 2001), and botulinum toxin (Minton 1995). The way these proteins cross the endosomal membrane depends on conformational changes in their structure caused by acidification of the endosomal vesicle. These proteins insert directly into the membrane to somehow facilitate transport of the catalytic domain from the endosome lumen to the cytosol. All of these proteins must disrupt membrane integrity to induce disease.

Some protein toxins, however, are not structurally equipped to enter cellular membranes directly. This group has evolved to co-opt pre-existing mechanisms of protein transport to deliver their catalytic domains to the cytosol. They do this by moving from the cell surface all the way to the endoplasmic reticulum (ER) (Johannes and Goud 1998; Lencer and Tsai 2003; Lord and Roberts 1998; Sandvig and van Deurs 2002). These toxins enter the ER as stably folded proteins and they co-opt components of ERAD to unfold and retro-translocate a portion of the toxin to the cytosol, presumably without disruption of membrane structure. Examples of these toxins are cholera toxin (CT) (De 1959; Dutta et al. 1959), Shiga toxin and the Shiga-like toxins (Stx) (O'Brien and Holmes 1987), *Pseudomonas* exotoxin A (ExoA) (Iglewski and Kabat 1975), and the plant toxin ricin (Balint 1974). They represent powerful model systems to study basic mechanisms of ERAD. Here, we will review the recent studies that define the molecular mechanisms for retro-translocation of CT and ricin to the cytosol. We will also briefly discuss Stx and ExoA, which are toxins that also enter the cytosol through the ER but the mechanisms are not as well defined.

2 Toxin Structure and Function

CT, Stx, ExoA and ricin are A/B-subunit toxins. They are composed of a catalytic polypeptide (the A subunit) that associates with one or more cell-binding peptides (the B subunit). Only the A-subunit crosses the limiting membrane of the ER to enter the cytosol. Ricin and ExoA are single A and B polypeptide toxins (A-B toxins), whereas CT and the Stx family consist of a single A chain associated with a pentamer of B chains (A-B_5 toxins) (Fig. 1). All these toxins are synthesized in pro-form and require activation by proteolytic cleavage in order to release the A-subunit from its A-B precursor or to cleave a precursor A polypeptide into A1- and A2-chains. The cleaved A and B, or the A1- and A2-, chains remain disulfide bonded.

2.1 CT Structure and Function

The CT A and B subunits are synthesized by the *Vibrio* with an N-terminal signal sequence directing them to the periplasm where holotoxin assembly occurs (Hirst and Holmgren 1987; Mekalanos et al. 1983). Five identical 11-kDa peptides associate in a ring-like structure to form the pentameric B-subunit (≈55 kDa) (Merritt and Hol 1995). The B-subunit is a lectin that binds to the oligosaccharide domain of a membrane glycolipid, ganglioside GM1. It

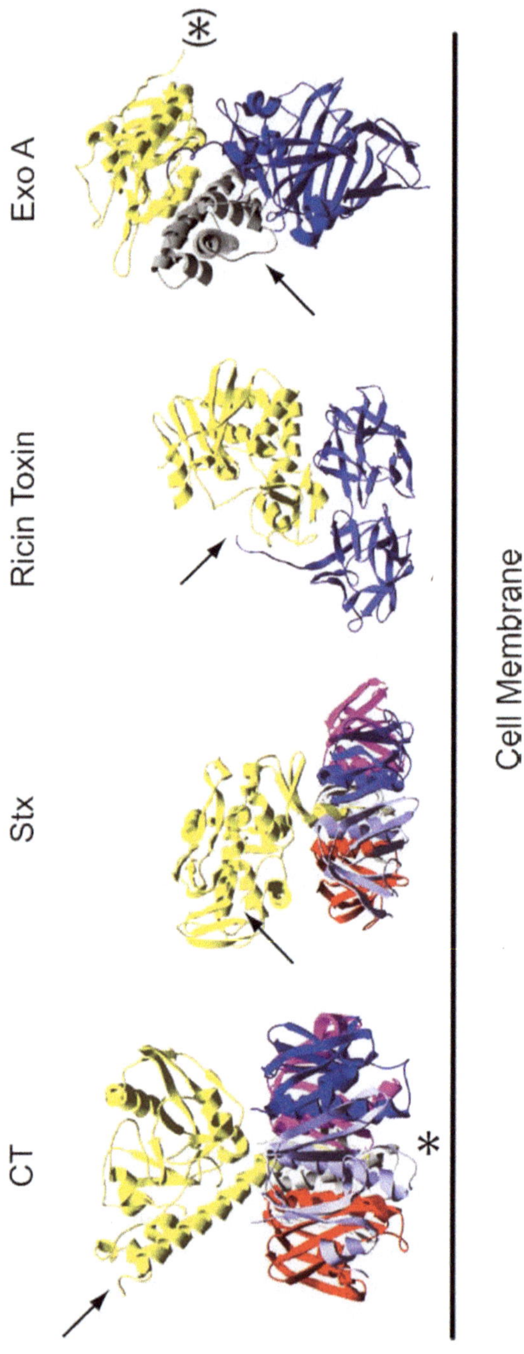
CT
Stx
Ricin Toxin
Exo A
(*)
*
Cell Membrane

Fig. 1 Crystal structure of toxins that retro-translocate from the ER. Shown from the *left* are the A-B_5 toxins cholera toxin (*CT*) and Shiga toxin (*Stx*), and the A-B toxins ricin and *Pseudomonas* exotoxin A (*ExoA*). Structures are shown sideways with the surface of the B-subunits containing the binding sites for their membrane receptors facing downwards. The toxin A subunits are shown in *yellow*. CT, Stx and ExoA have a proteolytic cleavage site within a loop subtended by a disulfide bond. *Arrows* indicate the position of the proteolytic cleavage site, and the *asterisks* indicate the positions of the Lys-Asp-Glu-Leu (KDEL) ER retrieval motif at the C-termini of the CT A2 and the ExoA catalytic domains, respectively. Ricin and Stx do not have KDEL motifs

can bind up to five gangliosides. The A-subunit assembles noncovalently with the B-subunit. Structurally and functionally, CT is very closely related to the heat-labile enterotoxins produced by *Escherichia coli* (Sixma et al. 1991, 1993; Spangler 1992). After secretion from the *Vibrio*, it is cleaved into A1- and A2-chains, which are still linked by a disulfide bond and extensive noncovalent interactions (Sixma et al. 1993). Even after reduction and proteolytic cleavage in vitro, the A1- and A2-chains remains stably associated. The A2-chain protrudes with its C-terminus through the central pore in the B-ring and tethers the A- and B-subunits together. The extreme C-terminus of the A2-chain has an ER-sorting KDEL motif facing the membrane (Lencer et al. 1995) (Other bacterial toxins also contain C-terminal ER retrieval signals, including *E. coli* heat labile enterotoxin [Lencer et al. 1995] and the protein synthesis inhibitor ExoA [Chaudary et al. 1990]). The A1-chain enters the cytosol and causes disease. It is an ADP-ribosyltransferase that modifies the heterotrimeric G protein Gs-α to activate adenylyl cyclase (Moss and Vaughan 1977). This induces intestinal chloride secretion that causes the massive secretory diarrhea seen in cholera (Kaper et al. 1995).

2.2 Ricin Structure and Function

Ricin contains a ribosome-inactivating A-chain (RTA) disulfide linked to a galactose-binding lectin (RTB). It is made in the producing castor oil plant as a precursor (proricin) in which a short linker separates the A- and B-chains (Lamb et al. 1985). The linker is a targeting signal that directs the transport of proricin to vacuoles (Frigerio et al. 2001), where proteolytic activation occurs to generate mature holotoxin (Harley and Lord 1985). Unlike CT, proteolytic cleavage and reduction of ricin in vitro causes complete dissociation of the two subunits. RTA enters the cytosol and cleaves a specific adenine residue from a highly conserved loop in the large rRNA of eukaryotic ribosomes (Endo et al. 1987). Ribosomes containing depurinated 28S rRNA cannot bind elongation factors and are therefore incapable of protein synthesis. While members of the bacterial STx family differ structurally from the plant toxin ricin in that they are A-B_5 toxins like CT, their catalytic A1 acts in an identical way to RTA, removing the same adenine residue from 28S rRNA (Endo et al. 1988). Although the primary sequence homology between the plant ricin and bacterial STx A-chains is not strong, the key catalytic residues are absolutely conserved. The A subunit of STx is cleaved into A1 and A2 akin to the processing of CT-A. Cleavage occurs after entry into the target cell and is mediated by the ubiquitous membrane-bound protease furin (Garred et al. 1995). In contrast to CT, however, the Stx A2 peptides do not possess a C-terminal ER retrieval signal.

Like ricin, ExoA is also synthesized as a single polypeptide chain and proteolytically cleaved and reduced to form the active toxin (McKee and FitzGerald 1999). Unlike ricin, it has an ER-targeting C-terminal KDEL-like motif (Chaudary et al. 1990).

3
The Pathway from Cell Surface to Endoplasmic Reticulum

3.1
Surface Binding and Cell Entry

Both CT and ricin begin their journey into the cell by binding to membrane receptors via the B-chain(s) followed by endocytosis (Lencer and Tsai 2003; Lord and Roberts 1998). CT binds specifically to the ganglioside GM1 (Holmgren et al. 1975), and ricin binds specifically to galactosides with a β1–4 linkage (Olsnes and Pihl 1982). Since a wide range of surface glycoproteins and glycolipids contain this galactoside, ricin is promiscuous in its binding to cell surface components. Like CT, the Stx family binds to a specific membrane glycolipid, interacting with the trisaccharide domain of globotriaosylceramide (Gb_3/CD77) (Lindberg et al. 1987). ExoA binds specifically to the α_2-macroglobulin receptor, which is a membrane protein (Kounnas et al. 1992).

Binding to these receptors is required for endocytosis that may occur by multiple mechanisms (Sandvig et al. 2004). Even so, each of these toxins can eventually be found in the early and recycling endosome (EE/RE) (Sandvig et al. 2004). While some toxins (diphtheria and anthrax toxins, for example) can cross into the cytosol from this endosomal compartment, others cannot (ricin, CT, Stx family and ExoA). This group must undergo retrograde transport to a location in the cell where protein translocation channels already exist – the endoplasmic reticulum (ER).

3.2
Transport to the Endoplasmic Reticulum

The pathway backwards from cell surface to ER for CT and the Stx family is dependent on lipid transport. Toxin binding to glycolipids with strong affinity for lipid microdomains (detergent-resistant membranes or "lipid rafts") appears to be critical for sorting into this pathway, not necessarily for a specific mechanism of internalization per se, but possibly for a subsequent sorting step at the level of EE/RE (Falguieres et al. 2001; Fujinaga et al. 2003). The lipid pathway may move directly from the *trans*-Golgi network (TGN) to ER without passing through the Golgi cisternae (Feng et al. 2004) and independently of COP-I vesicles that typify retrograde transport in the classic secretory pathway (Cosson and Letourneur 1994; Letourner et al. 1994).

ExoA also traffics backwards from the PM to the ER, but this appears to be critically dependent on binding to the KDEL-receptor that cycles between TGN, Golgi cisternae and ER (Miesenbock and Rothman 1995). It is proposed that the KDEL-receptor sorts ExoA in retrograde transport vesicles from TGN through the Golgi cisternae to the ER (Chaudary et al. 1990; Jackson et al. 1999; Kreitman and Pastan 1995; Seetharam et al. 1991).

The pathway backwards from PM to ER for ricin is not known, but is likely to follow either the lipid-dependent or KDEL-dependent pathways, or both. Ricin toxin does not contain an ER-sorting KDEL-motif and it can bind glycolipids that contain terminal galactose. There is also evidence that ricin may interact with the chaperone calreticulin in the Golgi complex. Calreticulin has a KDEL-motif, and it is possible that ricin co-opts calreticulin to sort backwards in the COP-I pathway to the ER by indirectly binding the KDEL-receptor (Day et al. 2001). The precise Golgi-to-ER pathway exploited by ricin remains unclear, however, since this toxin can enter cells inhibited in both the classical COP1-dependent and lipid-dependent pathways (Chen et al. 2003). In some situations, it also appears that ricin, like CT, can bypass the Golgi stack altogether (Llorente et al. 2003).

The emerging picture implies there are at least two routes available for toxins to progress from the TGN to ER. In one, the toxin, carried by the KDEL receptor or some other cycling protein receptor, moves via the Golgi complex in a COP1-dependent manner. This appears to be the route followed by PETx. In the second, the toxin is transported directly from the TGN to ER, bypassing the Golgi complex altogether, perhaps by virtue of an interaction with a raft-associated glycolipid receptor. This appears to be the pathway taken by CTx and STx (Fig. 2). Once in the ER, it is known, at least for CT, that the toxins bound to their lipid receptors can move anterograde from the ER to the cis Golgi and recycle back to the ER in a KDEL-mediated retrieval step. This

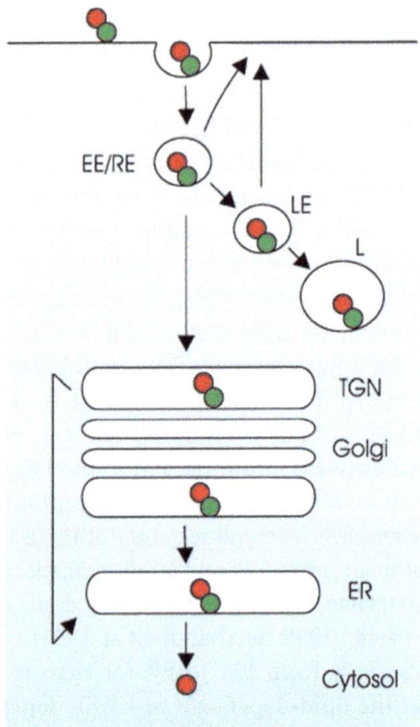

Fig. 2 The retrograde intracellular transport route from the cell surface to the ER. Surface-bound toxin enters the cell by endocytosis and is initially delivered to early/recycling endosomes (*EE/RE*), from where it moves directly to the *trans*-Golgi network (*TGN*). Nonproductive routing is thought to include transport to late endosomes (*LE*) and lysosomes (*L*), and recycling back to the plasma membrane (*PM*). Transport from the TGN to the ER may occur via the Golgi stack or may proceed directly

recycling for CT by binding to the KDEL receptor is proposed to retain CT in the ER and maximize retro-translocation of the A1-chain to the cytosol.

4
Retro-translocation

As described in detail elsewhere in this volume, the ER is a compartment in which an elaborate array of chaperones, enzymes and oxidoreductases functions to ensure the correct folding, assembly and disulfide bond formation of proteins entering the secretory pathway. It also operates a stringent quality control system, known as ER-associated protein degradation (ERAD), to retain proteins that fail to fold or to assemble properly. For such proteins, this system can facilitate their structural maturation; or if they are terminally misfolded, the ER can dispose of these proteins by retro-translocating them to the cytosol for degradation by the proteosome. It is this aspect of protein quality control in the biosynthetic pathway that CT, the Stx family, ricin, and ExoA co-opt to reach the cytosol and cause disease. Evidence for a functional correlation between ERAD and sensitivity to CT, ExoA and ricin has been provided in studies using mutant CHO cells that display either decreased or increased ERAD activities (Teter and Holmes 2002; Teter et al. 2003).

CT, and probably all the other toxins, enters the ER in its native A-B_5 (or A-B for ricin and ExoA) conformations (Fujinaga et al. 2003). Here, the toxins are first recognized, reduced, and unfolded by ER chaperones, and subsequently targeted to a protein-conducting channel for retro-translocation to the ER.

4.1 Substrate Recognition

One of the conundrums in the biology of the toxins that use the ER for retro-translocation is that the mammalian ER is a protein-folding environment functionally similar to both the bacterial periplasm and plant cell ER where the toxins are initially produced. In one instance, the toxins are folded and specific SH groups are oxidized to form disulfide bonds, and in the other instance, the toxins are unfolded and disulfide bonds reduced. In the case of CT, the explanation for this problem is known. The structure of CT contains a molecular switch that allows for toxin folding and assembly in the bacterial periplasm, but signals entry into the ERAD pathway for unfolding and subunit dissociation in the ER of target mammalian cells. This molecular switch is the critical protease site in the loop connecting the A1- and A2-chains that allows for toxin activation (see above). Proteolytic cleavage of this loop converts CT into a substrate for protein disulfide isomerase (PDI). This ER chaperone unfolds and dissociates the A1 chain from the rest of the toxin in preparation for retro-translocation to the cytosol (Tsai et al. 2001). The exact motif that allows PDI to act on CT is not known, but it likely has to do with exposure of hydrophobic domains in the A1-chain that are buried in the uncleaved proform of the toxin.

Given that all the A/B toxins require proteolytic cleavage for activation, we believe these other toxins that use ERAD also use the proteolytic motif as a molecular switch for entry into the ERAD pathway. Proteolytic cleavage and activation of the toxins can occur before or after entry into the target cell. In the case of CT, the toxin is cleaved after secretion from the *Vibrio* into the intestinal lumen, and in the case of ricin, the toxin is cleaved after entry into the vacuole that contains digestive enzymes (Lord 1985a). In the case of ExoA and Shiga toxin (and also potentially true for CT), the toxins are activated during entry into target cells by endogenous proteases (Gordon and Leppla 1994; Lencer et al. 1997).

4.2
Reduction, Unfolding and Subunit Dissociation

Based on the structure of the Sec61 translocon (Clemons et al. 2004; van den Berg et al. 2004), the catalytic domains of the toxins are assumed to be reductively separated from the B/A2-B_5 subunit(s) and unfolded to some degree before passing through the Sec61 channel. For ricin, reduction appears to be catalyzed in the ER by PDI (Spooner et al. 2004). There is evidence that PDI may also specifically reduce the nicked CT-A1 chain and ExoA (McKee and FitzGerald 1999; Orlandi 1997), but we do not obtain the same results for PDI when studied in vitro with CT (Tsai et al. 2001). It is possible that several other components of the ER can perform this function, PDI predominating because of its abundance.

The unfolding reaction for CT has been studied. Here, we find that PDI acts as redox-dependent chaperone to unfold the CT A1 chain and dissociate it from the rest of the toxin (Tsai et al. 2001). In its reduced form, PDI binds and unfolds the A1 chain. It may also target the PDI-A1 complex to the ER membrane (Tsai and Rapoport 2002). At this site, the ER oxidase ERO1 catalyzes the oxidation of PDI, causing the release of the A1 peptide directly or indirectly to the retro-translocation machinery. PDI, however, may not be a general catalyst for unfolding peptides because there is evidence that not every polypeptide bound by PDI is driven by a redox cycle (Lumb and Bulleid 2002). In the case of CT, the data show clearly that PDI can bind, unfold, and after catalysis by ERO1, release the A1 chain under physiological redox conditions (Tsai and Rapoport 2002; Tsai et al. 2001). PDI may not be the only chaperone responsible for unfolding and dissociating the CT A1 chain. One recent study suggests that the ER chaperone BiP may also play a role (Winkeler et al. 2003).

Unlike CT, PDI does not appear to promote unfolding or release of the RTA from the holotoxin, even though PDI is probably responsible for reducing the peptide (Bellisola et al. 2004; Spooner et al. 2004). It is known that reduction and dissociation is required for retro-translocation of RTA, at least in plant cells (Frigerio et al. 1998). When RTA is transiently expressed as a secretory protein without RTB in tobacco protoplasts, the RTA appears to be able to retro-translocate to the cytosol because it causes a significant ribotoxicity. When RTA is co-expressed with RTB, however, this toxicity is essentially mitigated because the two chains assemble into holotoxin in the ER lumen. Thus only the reduced and free RTA is competent for retro-translocation, but unlike CT, the RTA is released from its B-subunit in its fully folded conformation (Wright and Robertus 1987).

So how does RTA unfold to move through the retro-translocation pore? When RTA separates from RTB, several residues or surfaces are exposed that can insert into artificial membranes and this may induce unfolding. Indeed, RTA can interact with negatively charged phospholipid vesicles in a way that triggers structural changes in the protein and membrane destabilization (Day et al. 2002). Thus, it is possible that in vivo, RTA might interact with negatively charged phosphatidylserine on the lumenal surface of the ER membrane to unfold. It is also possible that ER chaperones might recognize these newly exposed RTA domains and catalyze the unfolding reaction itself, in a way that is similar to that shown for CT. There are no studies to date on the mechanism of Stx or ExoA unfolding.

4.3
Retro-translocation: The Protein Conducting Channel and Driving Force

How do the toxins move across the ER membrane? There is some evidence that the Sec61 translocon may be the protein-conducting channel. While the data to date are suggestive, they are indirect and not fully conclusive. In addition, the discovery of Derlin-1 and its function in the retro-translocation of MHC class I in US11 transfected cells (Lilley and Ploegh 2004; Ye et al. 2004) cast further doubt on the idea that Sec61 represents the sole protein-conducting channel for retro-translocation. The possible role of Derlin-1 in retro-translocation of the toxins, however, has not yet been tested directly.

In the case of CT, evidence for the involvement of Sec61 comes from the in vitro expression of the CT A1-chain in ER-derived microsomes (Schmitz et al. 2000). Here, the A1-chain was co-immunoprecipitated with components of Sec61, suggesting the capture of a translocating intermediate. In the case of ricin, RTA was co-immunoprecipitated with Sec61α as it entered mammalian cells in culture (Wesche et al. 1999). Furthermore, when RTA was expressed in yeast mutants defective in Sec61 function, the rate of degradation for RTA was reduced, suggesting that the defect in Sec61 caused a block in export of RTA to the cytosol. There is also some evidence that ExoA can use the Sec61 complex for retro-translocation (Koopmann et al. 2000). Almost nothing is known, however, about how Stx reaches the cytosol from the ER, or about what happens at the molecular level to the toxin B-subunits after the A-chains are dissociated. In some cases, the B-chains are rapidly degraded (Spooner et al. 2004), and in others they appear to remain stable for prolonged periods, possibly inside the ER (reviewed in Smith et al. 2004).

The driving force for retro-translocation of any protein also remains unknown. Since almost all terminally misfolded proteins known to be retro-translocated are poly-ubiquitinated, one idea is that the covalently attached ubiquitin polypeptides may act as a plug, preventing the retro-translocating protein from backsliding and thus acting as a ratchet to drive the transport reaction. This idea has been tested for CT and ricin. The data show that a mutant CT A1 chain with its N-terminus chemically blocked and all lysines mutated to arginine and thus lacking all sites for poly-ubiquitination and a mutant RTA lacking all lysines remain essentially fully toxic (Deeks et al. 2002; Rodighiero et al. 2002). These data show that poly-ubiquitination cannot be only driving force for retro-translocation of the toxins. It is possible that the ability of the CT A1 chain to rapidly and spontaneously refold may cause the toxin to ratchet itself out of the retro-translocating pore, but such rapid refolding is not a characteristic of all the toxins and this idea has not been conclusively tested. The AAA-ATPase p97 and its adaptor molecules Ufd1 and Npl4 are involved in retro-translocation of other ERAD substrates (see, for example, Bays and Hampton 2002) and may be involved in toxin retro-translocation, but this also has not yet been tested.

4.4 After Retro-translocation: Refolding and Escape from Proteosomes

To act on their substrates, the retro-translocated toxins must to some extent disengage from the sequential steps leading to degradation that normally occurs upon extraction of ERAD substrates from the membrane. Assuming the toxins are unfolded for retro-translocation, they may achieve this by rapidly refolding in the cytosol and avoiding the proteosome, as has been demonstrated for the CT A1 chain. Alternatively, as first proposed by Hazes and Read (1997), they may simply avoid poly-ubiquitination after arrival in the cytosol because these toxins uniquely contain a paucity of lysines in their primary sequence. Only the primary amines contained in lysines and at the N-terminus of polypeptides are substrate for poly-ubiquitination.

Remarkably, the A-chains of seven different toxins believed to cross into the cytosol through the ER posses only 2.3 lysyl residues per polypeptide on average. This is in marked contrast to the structurally related but nontoxic type I ribosome-inactivating proteins (ricin A chain-like polypeptides without a B-chain) that contain an average of 18.6 lysyl residues (Deeks et al. 2002), and proteins that exit directly from the endosome; for example, diphtheria toxin A-chain contains 16 lysyl residues (Hazes and Read 1997). It has also been shown that the introduction of additional lysyl residues into both RTA and the CT A1 chain significantly increases their susceptibility to proteasomal

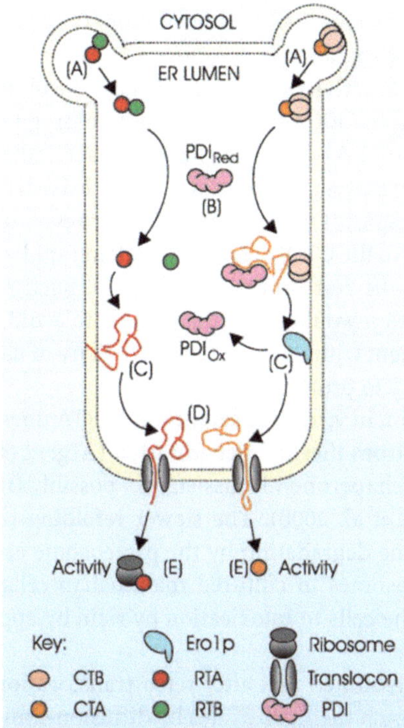

Fig. 3 The fate of CT and ricin upon reaching the ER lumen. (*A*) CT enters the ER as a fully assembled A-B_5 toxin but proteolytically cleaved in the A subunit, while ricin enters as an A-B heterodimer. (*B*) The reduced form of protein disulfide isomerase (*PDI*) interacts with both toxins but with different results. Reduced PDI recognizes cleaved CT, binds to and unfolds the A1 subunit, and dissociates it from the rest of the toxin. PDI reductively cleaves the disulfide bond linking the A- and B-subunits of ricin to release the free but still folded RTA subunit. (*C*) In the case of CT, the PDI-A1 complex is targeted to the luminal surface of the ER membrane where Ero1 oxidizes PDI to release the unfolded A1-subunit. RTA interacts with the ER membrane, which results in RTA unfolding. (*D*) The unfolded CT-A1 and RTA subunits are perceived as ERAD substrates and retro-translocated to the cytosol, possibly passing through the Sec61p translocon. (*E*) Once the cytosol is reached, the paucity of lysine residues in CT-A1 and RTA prevents extensive ubiquitination, allowing the toxin to uncouple from the proteasomal degradation step of ERAD. CT-A1 does this by rapidly refolding into its native conformation, while RTA appears to use the ribosome as a chaperone in an assisted refolding step. Once refolded, the toxin A subunits are proteolytically resistant and biologically active, and go on to modify their cytosolic substrates

degradation and reduces toxicity (Deeks et al. 2002; Rodighiero et al. 2002). These data support the idea that the low lysine content of the ER retro-translocating toxins has resulted from evolutionary pressure for the toxins to avoid proteasomal degradation.

To have enzymatic activity in the target cell cytosol, the toxins must also refold after crossing the ER membrane. The refolding reaction in vitro has been modeled for the CT A1-chain and for RTA. For CT, the unfolded A1-chain can be released from PDI after oxidation. Once released, the A1-chain refolds spontaneously and rapidly (within 5 s) (Rodighiero et al. 2002). This has led us to propose that in vivo the CT A1-chain may refold rapidly and spontaneously as it emerges from the retro-translocating pore. Such rapid refolding into its native conformation would cause the peptide to avoid degradation by the proteosome, consistent with the fact that the toxicity of native CT displays no detectable sensitivity to proteasome inhibition.

In the case of ricin, in vitro studies show that RTA does not spontaneously refold after release from thermal denaturation (Argent et al. 2000). Instead, RTA requires some chaperone-like assistance, possibly from the target ribosome itself (Argent et al. 2000). The slower refolding rate of RTA renders it susceptible to some degradation by the proteosome as evidenced in vivo. Inhibition of proteasomes in cultured mammalian cells, for example, will typically sensitize the cells to intoxication by ricin by approximately two- to threefold.

In all cases, it is assumed that after retro-translocation and refolding the catalytic A chains reach their substrates by diffusion-limited reactions. This, however, has not been studied in vivo and it is possible that the toxins exploit other cytosolic components. Our current understanding of the fate of CT and ricin after their delivery into the ER lumen is summarized in Fig. 3.

5 Toxin Retro-translocation in Plant Cells

An intriguing question surrounds the biosynthesis of ricin. How does the plant synthesize large amounts of this toxin when the producing ribosomes are sensitive to its action (Harley and Beevers 1982)? It has been speculated that since ricin is initially made as an ER-targeted proricin precursor (Butterworth and Lord 1983), the presence of RTA in this context somehow prevents its recognition as an ERAD substrate and therefore precludes its retro-translocation to the cytosol. There is some experimental support for this, since when tobacco protoplasts are forced to express RTA by itself, the polypeptide is found to become co-translationally glycosylated whilst remaining toxic to ribosomes

(Di Cola et al. 2001). Metabolic labelling showed that most RTA was degraded by proteasomes in a process preceded by a cytosolic deglycosylation step. This was the first reported example of a retro-translocation pathway in plant cells. Although degradation was the fate of most of the retro-translocated RTA, some of this enzyme was able to disengage from downstream steps and damage the tobacco ribosomes. This does not occur during ricin biosynthesis. In *Ricinus* seeds, ER-segregated proricin is transported via the secretory pathway to the protein storage vacuoles (Lord 1985a, 1985b). Here, the active holotoxin is generated by proteolysis of the 12 residues' vacuolar targeting signal that links the RTA and RTB sequences in the proricin precursor (Harley and Lord 1985). Since the ribosomes of *Ricinus* endosperm cells show no sign of damage while synthesizing and storing large amounts of active toxin, ricin clearly does not retro-translocate across the vacuolar membrane. This strategy of synthesis allows the seeds to make and store a highly potent toxin to 5% of total particulate protein without compromising survival.

6
Concluding Remarks

It is now clear that certain toxins move into the cytosol of target cells by co-opting components of ERAD. The molecular components of this pathway are only now emerging. Many questions remain unanswered and we look forward to future studies that examine the biology of these remarkable proteins.

Acknowledgements Work at Warwick is supported by the UK Biotechnology and Biological Sciences Research Council and a Wellcome Trust Programme grant (063058/Z/00/Z). Dr. Lencer is supported by research grants DK48106, DK57827 and DK34854.

References

Argent RH, Parrott AM, Day PJ, Roberts LM, Stockley PG, Lord JM, Radford SE (2000) Ribosome-mediated folding of partially unfolded ricin A-chain. J Biol Chem 275:9263–9269

Balint G (1974) Ricin: the toxic protein of castor oil seeds. Toxicology 2:77–102

Bays N, Hampton R (2002) Cdc48-Ufd1-Npl4: stuck in the middle with Ub. Curr Biol 12:R366–R371

Bellisola G, Fracasso G, Ippoliti R, Menestrina G, Rosen A, Solda S, Udali S, Tomazzolli R, Tridente G, Colombatti M (2004) Reductive activation of ricin and ricin A-chain immunotoxins by protein disulfide isomerase and thioredoxin reductase. Biochem Pharmacol 67:1721–1731

Bhakdi S, Tranum-Jensen J (1991) Alpha-toxin of Staphylococcus aureus. Microbiol Rev 55:733–751

Buckley J, Halasa L, Lund K, MacIntyre S (1981) Purification and some properties of the hemolytic toxin aerolysin. Can J Biochem 59:430–436

Butterworth AG, Lord JM (1983) Ricin and Ricinus communis agglutinin subunits are all derived from a single-size polypeptide precursor. Eur J Biochem 137:57–65

Chaudhary V, Jinno Y, FitzGerald D, Pastan I (1990) Pseudomonas exotoxin contains a specific sequence at the carboxyl terminus that is required for cytotoxicity. Proc Natl Acad Sci U S A 87:308–312

Chen A, Abu Jarour RJ, Draper RK (2003) Evidence that the transport of ricin to the cytoplasm is independent of both Rab6A and COPI J Cell Sci 116:3503–3510

Clemons W, Menetret JF, Akey CW, Rapoport T (2004) Structural insight into the protein translocation channel. Curr Opin Struct Biol 14:390–396

Cosson P, Letourneur F (1994) Coatomer interaction with di-lysine endoplasmic reticulum retention motifs. Science 263:1629–1631

Day PJ, Owens SR, Wesche J, Olsnes S, Roberts LM, Lord JM (2001) An interaction between ricin and calreticulin that may have implications for toxin trafficking. J Biol Chem 276:7202–7208

Day PJ, Pinheiro TJ, Roberts LM, Lord JM (2002) Binding of ricin A-chain to negatively charged phospholipid vesicles leads to protein structural changes and destabilizes the lipid bilayer. Biochemistry 41:2836–2843

De S (1959) Enterotoxicity of bacteria-free culture filtrate of Vibrio cholerae. Nature 183:1533–1534

Deeks ED, Cook JP, Day PJ, Smith DC, Roberts LM, Lord JM (2002) The low lysine content of ricin A chain reduces the risk of proteolytic degradation after translocation from the endoplasmic reticulum to the cytosol. Biochemistry 41:3405–3413

Di Cola A, Frigerio L, Lord JM, Ceriotti A, Roberts LM (2001) Ricin A chain without its partner B chain is degraded after retrotranslocation from the endoplasmic reticulum to the cytosol in plant cells. Proc Natl Acad Sci U S A 98:14726–14731

Dutta N, Panse M, Kulkarni D (1959) Role of cholera toxin in experimental cholera. J Bacteriol 78:594–595

Endo Y, Mitsui K, Motizuki M, Tsurugi K (1987) The mechanism of action of ricin and related toxic lectins on eukaryotic ribosomes. The site and the characteristics of the modification in 28 S ribosomal RNA caused by the toxins. J Biol Chem 262:5908–5912

Endo Y, Tsurugi K, Yutsudo T, Takeda Y, Ogasawara T, Igarashi K (1988) Site of action of a Vero toxin (VT2) from Escherichia coli O157:H7 and of Shiga toxin on eukaryotic ribosomes. RNAN-glycosidase activity of the toxins. Eur J Biochem 171:25–50

Falguieres T, Mallard F, Baron C, Hanau D, Lingwood C, Goud B, Salamero J, Johannes L (2001) Targeting of Shiga toxin B-subunit to retrograde transport route in association with detergent-resistant membranes. Mol Biol Cell 12:2453–2468

Feng Y, Jadhav A, Rodighiero C, Fujinaga Y, Kirchhausen T, Lencer W (2004) Retrograde transport of cholera toxin from the plasma membrane to the endoplasmic reticulum requires the trans-Golgi network but not the Golgi apparatus in Exo2-treated cells. EMBO Rep 5:596–601

Frigerio L, Jolliffe NA, Di Cola A, Felipe DH, Paris N, Neuhaus JM, Lord JM, Ceriotti A, Roberts LM (2001) The internal propeptide of the ricin precursor carries a sequence-specific determinant for vacuolar sorting. Plant Physiol 126:167–175

Frigerio L, Vitale A, Lord JM, Ceriotti A, Roberts LM (1998) Free ricin A chain, proricin, and native toxin have different cellular fates when expressed in tobacco protoplasts. J Biol Chem 273:14194–14199

Fujinaga Y, Wolf AA, Rodighiero C, Wheeler H, Tsai B, Allen L, Jobling MG, Rapoport T, Holmes RK, Lencer WI (2003) Gangliosides that associate with lipid rafts mediate transport of cholera and related toxins from the plasma membrane to endoplasmic reticulum. Mol Biol Cell 14:4783–4793

Garred O, van Deurs B, Sandvig K (1995) Furin-induced cleavage and activation of Shiga toxin. J Biol Chem 270:10817–10821

Gordon V, Leppla S (1994) Protoelytic activation of bacterial toxins: role of bacterial and host cell proteases. Infect Immun 62:333–340

Harley SM, Beevers H (1982) Ricin inhibition of in vivo protein synthesis by plant ribosomes. Proc Natl Acad Sci U S A 79:5935–5938

Harley SM, Lord JM (1985) In vitro endoproteolytic cleavage of castor bean lectin precursors. Plant Sci 41:111–116

Hazes B, Read RJ (1997) Accumulating evidence suggests that several AB-toxins subvert the endoplasmic reticulum-associated protein degradation pathway to enter target cells. Biochemistry 36:11051–11054

Hirst TR, Holmgren J (1987) Transient entry of enterotoxin subunits into the periplasm occurs during their secretion from Vibrio cholerae. J Bacteriol 169:1037–1045

Holmgren J, Lonnroth I, Mansson J, Svennerholm L (1975) Interaction of cholera toxin and membrane GM1 ganglioside of small intestine. Proc Natl Acad Sci U S A 72:2520–2524

Iglewski B, Kabat D (1975) NAD-dependent inhibition of protein synthesis by Pseudomonas aeruginosa toxin. Proc Natl Acad Sci U S A 72:2284–2288

Jackson ME, Simpson JC, Girod A, Pepperkok R, Roberts LM, Lord JM (1999) The KDEL retrieval system is exploited by Pseudomonas exotoxin A, but not by Shiga-like toxin-1, during retrograde transport from the Golgi complex to the endoplasmic reticulum. J Cell Sci 112:467–475

Johannes L, Goud B (1998) Surfing on a retrograde wave: how does Shiga toxin reach the endoplasmic reticulum? Trends Cell Biol 8:158–162

Kaper J, Morris J, Levine M (1995) Cholera. Clin MicrobiolRev 8:48–86

Koopmann J O, Albring J, Huter E, Bulbuc N, Spee P, Neefjes J, Hammerling GJ, Momburg F (2000) Export of antigenic peptides from the endoplasmic reticulum intersects with retrograde protein translocation through the Sec61p channel. Immunity 13:117–127

Kounnas M, Morris R, Thompson M, FitzGerald D, Strickland D, Saelinger C (1992) The alpha2-macroglobulin receptor/low density lipoprotein-related protein binds and internalizes Pseudomonas exotoxin A. J Biol Chem 267:12420–12423

Kreitman R, Pastan I (1995) Importance of the glutamate residue of the KDEL in increasing the cytotoxicity of Pseudomonas exotoxin derivatives and for increased binding to the KDEL receptor. Biochem J 307:29–37

Lamb FI, Roberts LM, Lord JM (1985) Nucleotide sequence of cloned cDNA coding for preproricin. Eur J Biochem 148:265–270

Lencer WI, Constable C, Moe S, Jobling MG, Webb HM, Ruston S, Madara JL, Hirst TR, Holmes RK (1995) Targeting of cholera toxin and Escherichia coli heat labile toxin in polarized epithelia: role of COOH-terminal KDEL. J Cell Biol 131:951–962

Lencer WI, Constable C, Moe S, Rufo PA, Wolf A, Jobling MG, Ruston SP, Madara JL, Holmes RK, Hirst TR (1997) Proteolytic activation of cholera toxin and Escherichia coli labile toxin by entry into host epithelial cells. Signal transduction by a protease-resistant toxin variant. J Biol Chem 272:15562–15568

Lencer WI, Tsai B (2003) The intracellular voyage of cholera toxin: going retro. Trends Biochem Sci 28:639–645

Letourner F, Gaynor E, Hennecke S, Demolliere C, Duden R, Emr S, Riezman H, Cosson P (1994) Coatomer is essential for retrieval of dilysine-tagged proteins to the endoplasmic reticulum. Cell 79:1199–1207

Lilley BN, Ploegh HL (2004) A membrane protein required for dislocation of misfolded proteins from the ER. Nature 429:834–840

Lindberg A, Brown J, Stromberg N, Westling-Ryd M, Schultz J, Karlsson K (1987) Identification of the carbohydrate receptor for Shiga toxin produced by Shigella dysenteriae type 1. J Biol Chem 262:1779–1785

Llorente A, Lauvrak SU, Van Deurs B, Sandvig K (2003) Induction of direct endosome to endoplasmic reticulum transport in Chinese Hamster Ovary (CHO) cells (LdlF) with a temperature-sensitive defect in epsilon-coatomer protein (epsilon-COP). J Biol Chem 278:35850–35855

Lord JM (1985a) Precursors of ricin and Ricinus communis agglutinin. Glycosylation and processing during synthesis and intracellular transport. Eur J Biochem 146:411–416

Lord JM (1985b) Synthesis and intracellular transport of lectin and storage protein precursors in endosperm from castor bean. Eur J Biochem 146:403–409

Lord JM, Roberts LM (1998) Toxin entry: retrograde transport through the secretory pathway. J Cell Biol 140:733–736

Lumb RA, Bulleid NJ (2002) Is protein disulfide isomerase a redox-dependent molecular chaperone? EMBO J 21:6763–6770

McKee ML, FitzGerald DJ (1999) Reduction of furin-nicked Pseudomonas exotoxin A: an unfolding story. Biochemistry 38:16507–16513

Mekalanos J, Swartz D, Pearson G, Harford N, Groyne F, de Wilde M (1983) Cholera toxin genes: nucleotide sequence, deletion analysis and vaccine development. Nature 306:551–557

Merritt E, Hol W (1995) AB5 toxins. Curr Opin Struct Biol 5:165–171

Miesenbock G, Rothman J (1995) The capacity to retrieve escaped ER proteins extends to the trans-most cisterna of the Golgi stack. J Cell Biol 129:309–319

Minton N (1995) Molecular genetics of clostridial neurotoxins. Curr Top Microbiol Immunol 195:161–194

Mock M, Fouet A (2001) Anthrax. Annu Rev Microbiol 55:647–671

Moss J, Vaughan M (1977) Mechanism of action of choleragen: evidence for ADP-ribosyltransferase activity with arginine as an acceptor. J Biol Chem 252:2455–2457

O'Brien A, Holmes R (1987) Shiga and the Shiga-like toxins. Microbiol Rev 51:206–220

Olsnes S, Pihl A (1982) Toxic lectins and related proteins. In: Cohen P, van Heyningen S (eds) Molecular action of toxins and viruses. Elsevier, Amsterdam, pp 51–105
Orlandi PA (1997) Protein-disulfide isomerase-mediated reduction of the A subunit of cholera toxin in a human intestinal cell line. J Biol Chem 272:4591–4599
Pappenheimer AM Jr (1977) Diphtheria toxin. Annu Rev Biochem 46:29–94
Rodighiero C, Tsai B, Rapoport TA, Lencer WI (2002) Role of ubiquitination in retro-translocation of cholera toxin and escape of cytosolic degradation. EMBO Rep 3:1222–1227
Sandvig K, Spilsberg B, Lauvrak SU, Torgersen ML, Iversen TG, van Deurs B (2004) Pathways followed by protein toxins into cells. Int J Med Microbiol 293:483–490
Sandvig K, van Deurs B (2002) Transport of protein toxins into cells: pathways used by ricin, cholera toxin and Shiga toxin. FEBS Lett 529:29–53
Schmitz A, Herrgen H, Winkeler A, Herzog V (2000) Cholera toxin is exported from microsomes by the Sec61p complex. J Cell Biol 148:1203–1212
Seetharam S, Chaudhary VK, FitzGerald D, Pastan I (1991) Increased cytotoxic activity of Pseudomonas exotoxin and two chimeric toxins ending in KDEL. J Biol Chem 266:17376–17381
Sekura R, Moss J, Vaughan M (1985) Pertussis toxin. Academic Press, Orlando
Sixma T, Kalk K, van Zanten B, Dauter Z, Kingma J, Witholt B, Hol W (1993) Refined structure of Escherichia coli heat-labile enterotoxin, a close relative of cholera toxin. J Mol Biol 230:890–918
Sixma T, Pronk S, Kalk K, Wartna E, van Zanten B, Witholt B, Hol W (1991) Crystal structure of a cholera toxin-related heat-labile enterotoxin from E coli. Nature 351:371–377
Smith DC, Lord JM, Roberts LM, Johannes L (2004) Glycosphingolipids as toxin receptors. Semin Cell Dev Biol 15:497–408
Spangler B (1992) Structure and function of cholera toxin and the related Escherichia coli heat-labile enterotoxin. Microbiol Rev 56:622–647
Spooner RA, Watson PD, Marsden CJ, Smith DC, Moore KA, Cook J P, Lord JM, Roberts LM (2004) Protein disulphide isomerase reduces ricin to its A and B chains in the endoplasmic reticulum. Biochem J 383:285–293
Teter K, Holmes RK (2002) Inhibition of endoplasmic reticulum-associated degradation in CHO cells resistant to cholera toxin, Pseudomonas aeruginosa exotoxin A, and ricin. Infect Immun 70:6172–6179
Teter K, Jobling MG, Holmes RK (2003) A class of mutant CHO cells resistant to cholera toxin rapidly degrades the catalytic polypeptide of cholera toxin and exhibits increased endoplasmic reticulum-associated degradation. Traffic 4:232–242
Tsai B, Rapoport TA (2002) Unfolded cholera toxin is transferred to the ER membrane and released from protein disulfide isomerase upon oxidation by Ero1. J Cell Biol 159:207–216
Tsai B, Rodighiero C, Lencer WI, Rapoport TA (2001) Protein disulfide isomerase acts as a redox-dependent chaperone to unfold cholera toxin. Cell 104:937–948
Van den Berg B, Clemons WM Jr, Collinson I, Modis Y, Hartmann E, Harrison SC, Rapoport T (2004) X-ray structure of a protein-conducting channel. Nature 427:26–44

Wesche J, Rapak A, Olsnes S (1999) Dependence of ricin toxicity on translocation of the toxin A-chain from the endoplasmic reticulum to the cytosol. J Biol Chem 274:34443–34449

Winkeler A, Godderz D, Herzog V, Schmitz A (2003) BiP-dependent export of cholera toxin from endoplasmic reticulum-derived microsomes. FEBS Lett 554:439–442

Wright HT, Robertus JD (1987) The intersubunit disulfide bridge of ricin is essential for cytotoxicity. Arch Biochem Biophys 256:280–284

Ye Y, Shibata Y, Yun C, Ron D, Rapoport TA (2004) A membrane protein complex mediates retro-translocation from the ER lumen into the cytosol. Nature 429:841–847

Subject Index

Current Topics in Microbiology and Immunology

Volumes published since 1989 (and still available)

Vol. 256: **Schmaljohn, Connie S.; Nichol, Stuart T. (Eds.):** Hantaviruses. 2001. 24 figs. XI, 196 pp. ISBN 3-540-41045-7

Vol. 257: **van der Goot, Gisou (Ed.):** PoreForming Toxins, 2001. 19 figs. IX, 166 pp. ISBN 3-540-41386-3

Vol. 258: **Takada, Kenzo (Ed.):** Epstein-Barr Virus and Human Cancer. 2001. 38 figs. IX, 233 pp. ISBN 3-540-41506-8

Vol. 259: **Hauber, Joachim, Vogt, Peter K. (Eds.):** Nuclear Export of Viral RNAs. 2001. 19 figs. IX, 131 pp. ISBN 3-540-41278-6

Vol. 260: **Burton, Didier R. (Ed.):** Antibodies in Viral Infection. 2001. 51 figs. IX, 309 pp. ISBN 3-540-41611-0

Vol. 261: **Trono, Didier (Ed.):** Lentiviral Vectors. 2002. 32 figs. X, 258 pp. ISBN 3-540-42190-4

Vol. 262: **Oldstone, Michael B.A. (Ed.):** Arenaviruses I. 2002. 30 figs. XVIII, 197 pp. ISBN 3-540-42244-7

Vol. 263: **Oldstone, Michael B. A. (Ed.):** Arenaviruses II. 2002. 49 figs. XVIII, 268 pp. ISBN 3-540-42705-8

Vol. 264/I: **Hacker, Jörg; Kaper, James B. (Eds.):** Pathogenicity Islands and the Evolution of Microbes. 2002. 34 figs. XVIII, 232 pp. ISBN 3-540-42681-7

Vol. 264/II: **Hacker, Jörg; Kaper, James B. (Eds.):** Pathogenicity Islands and the Evolution of Microbes. 2002. 24 figs. XVIII, 228 pp. ISBN 3-540-42682-5

Vol. 265: **Dietzschold, Bernhard; Richt, Jürgen A. (Eds.):** Protective and Pathological Immune Responses in the CNS. 2002. 21 figs. X, 278 pp. ISBN 3-540-42668X

Vol. 266: **Cooper, Koproski (Eds.):** The Interface Between Innate and Acquired Immunity, 2002. 15 figs. XIV, 116 pp. ISBN 3-540-42894-X

Vol. 267: **Mackenzie, John S.; Barrett, Alan D. T.; Deubel, Vincent (Eds.):** Japanese Encephalitis and West Nile Viruses. 2002. 66 figs. X, 418 pp. ISBN 3-540-42783X

Vol. 268: **Zwickl, Peter; Baumeister, Wolfgang (Eds.):** The Proteasome-Ubiquitin Protein Degradation Pathway. 2002. 17 figs. X, 213 pp. ISBN 3-540-43096-2

Vol. 269: **Koszinowski, Ulrich H.; Hengel, Hartmut (Eds.):** Viral Proteins Counteracting Host Defenses. 2002. 47 figs. XII, 325 pp. ISBN 3-540-43261-2

Vol. 270: **Beutler, Bruce; Wagner, Hermann (Eds.):** Toll-Like Receptor Family Members and Their Ligands. 2002. 31 figs. X, 192 pp. ISBN 3-540-43560-3

Vol. 271: **Koehler, Theresa M. (Ed.):** Anthrax. 2002. 14 figs. X, 169 pp. ISBN 3-540-43497-6

Vol. 272: **Doerfler, Walter; Böhm, Petra (Eds.):** Adenoviruses: Model and Vectors in Virus-Host Interactions. Virion and Structure, Viral Replication, Host Cell Interactions. 2003. 63 figs., approx. 280 pp. ISBN 3-540-00154-9

Vol. 273: **Doerfler, Walter; Böhm, Petra (Eds.):** Adenoviruses: Model and Vectors in VirusHost Interactions. Immune System, Oncogenesis, Gene Therapy. 2004. 35 figs., approx. 280 pp. ISBN 3-540-06851-1

Vol. 274: **Workman, Jerry L. (Ed.):** Protein Complexes that Modify Chromatin. 2003. 38 figs., XII, 296 pp. ISBN 3-540-44208-1

Vol. 275: **Fan, Hung (Ed.):** Jaagsiekte Sheep Retrovirus and Lung Cancer. 2003. 63 figs., XII, 252 pp. ISBN 3-540-44096-3

Vol. 276: **Steinkasserer, Alexander (Ed.):** Dendritic Cells and Virus Infection. 2003. 24 figs., X, 296 pp. ISBN 3-540-44290-1

Vol. 277: **Rethwilm, Axel (Ed.):** Foamy Viruses. 2003. 40 figs., X, 214 pp. ISBN 3-540-44388-6

Vol. 278: **Salomon, Daniel R.; Wilson, Carolyn (Eds.):** Xenotransplantation. 2003. 22 figs., IX, 254 pp. ISBN 3-540-00210-3

Vol. 279: **Thomas, George; Sabatini, David; Hall, Michael N. (Eds.):** TOR. 2004. 49 figs., X, 364 pp. ISBN 3-540-00534X

Vol. 280: **Heber-Katz, Ellen (Ed.):** Regeneration: Stem Cells and Beyond. 2004. 42 figs., XII, 194 pp. ISBN 3-540-02238-4

Vol. 281: **Young, John A. T. (Ed.):** Cellular Factors Involved in Early Steps of Retroviral Replication. 2003. 21 figs., IX, 240 pp. ISBN 3-540-00844-6

Vol. 282: **Stenmark, Harald (Ed.):** Phosphoinositides in Subcellular Targeting and Enzyme Activation. 2003. 20 figs., X, 210 pp. ISBN 3-540-00950-7

Vol. 283: **Kawaoka, Yoshihiro (Ed.):** Biology of Negative Strand RNA Viruses: The Power of Reverse Genetics. 2004. 24 figs., IX, 350 pp. ISBN 3-540-40661-1

Vol. 284: **Harris, David (Ed.):** Mad Cow Disease and Related Spongiform Encephalopathies. 2004. 34 figs., IX, 219 pp. ISBN 3-540-20107-6

Vol. 285: **Marsh, Mark (Ed.):** Membrane Trafficking in Viral Replication. 2004. 19 figs., IX, 259 pp. ISBN 3-540-21430-5

Vol. 286: **Madshus, Inger H. (Ed.):** Signalling from Internalized Growth Factor Receptors. 2004. 19 figs., IX, 187 pp. ISBN 3-540-21038-5

Vol. 287: **Enjuanes, Luis (Ed.):** Coronavirus Replication and Reverse Genetics. 2005. 49 figs., XI, 257 pp. ISBN 3-540-21494-1

Vol. 288: **Mahy, Brain W. J. (Ed.):** Foot-and-Mouth-Disease Virus. 2005. 16 figs., IX, 178 pp. ISBN 3-540-22419X

Vol. 289: **Griffin, Diane E. (Ed.):** Role of Apoptosis in Infection. 2005. 40 figs., IX, 294 pp. ISBN 3-540-23006-8

Vol. 290: **Singh, Harinder; Grosschedl, Rudolf (Eds.):** Molecular Analysis of B Lymphocyte Development and Activation. 2005. 28 figs., XI, 255 pp. ISBN 3-540-23090-4

Vol. 291: **Boquet, Patrice; Lemichez Emmanuel (Eds.)** Bacterial Virulence Factors and Rho GTPases. 2005. 28 figs., IX, 196 pp. ISBN 3-540-23865-4

Vol. 292: **Fu, Zhen F (Ed.):** The World of Rhabdoviruses. 2005. 27 figs., X, 210 pp. ISBN 3-540-24011-X

Vol. 293: **Kyewski, Bruno; Suri-Payer, Elisabeth (Eds.):** CD4+CD25+ Regulatory T Cells: Origin, Function and Therapeutic Potential. 2005. 22 figs., XII, 332 pp. ISBN 3-540-24444-1

Vol. 294: **Caligaris-Cappio, Federico, Dalla Favera, Ricardo (Eds.):** Chronic Lymphocytic Leukemia. 2005. 25 figs., VIII, 187 pp. ISBN 3-540-25279-7

Vol. 295: **Sullivan, David J.; Krishna Sanjeew (Eds.):** Malaria: Drugs, Disease and Post-genomic Biology. 2005. 40 figs., XI, 446 pp. ISBN 3-540-25363-7

Vol. 296: **Oldstone, Michael B. A. (Ed.):** Molecular Mimicry: Infection Induced Autoimmune Disease. 2005. 28 figs., VIII, 167 pp. ISBN 3-540-25597-4

Vol. 297: **Langhorne, Jean (Ed.):** Immunology and Immunopathogenesis of Malaria. 2005. 8 figs., XII, 236 pp. ISBN 3-540-25718-7

Vol. 298: **Vivier, Eric; Colonna, Marco (Eds.):** Immunobiology of Natural Killer Cell Receptors. 2006. 27 figs., VIII, 286 pp. ISBN 3-540-26083-8

Vol. 299: **Domingo, Esteban (Ed.):** Quasispecies: Concept and Implications for Virology. 2006. 44 figs., XII, 418 pp. ISBN 3-540-26395-0

MIX
Papier aus verantwortungsvollen Quellen
Paper from responsible sources
FSC® C105338

If you have any concerns about our products,
you can contact us on
ProductSafety@springernature.com

In case Publisher is established outside the EU,
the EU authorized representative is:
Springer Nature Customer Service Center GmbH
Europaplatz 3, 69115 Heidelberg, Germany

Printed by Libri Plureos GmbH
in Hamburg, Germany